Springer Theses

Recognizing Outstanding Ph.D. Research

More information about this series at http://www.springer.com/series/8790

Aims and Scope

The series "Springer Theses" brings together a selection of the very best Ph.D. theses from around the world and across the physical sciences. Nominated and endorsed by two recognized specialists, each published volume has been selected for its scientific excellence and the high impact of its contents for the pertinent field of research. For greater accessibility to non-specialists, the published versions include an extended introduction, as well as a foreword by the student's supervisor explaining the special relevance of the work for the field. As a whole, the series will provide a valuable resource both for newcomers to the research fields described, and for other scientists seeking detailed background information on special questions. Finally, it provides an accredited documentation of the valuable contributions made by today's younger generation of scientists.

Theses are accepted into the series by invited nomination only and must fulfill all of the following criteria

- They must be written in good English.
- The topic should fall within the confines of Chemistry, Physics, Earth Sciences, Engineering and related interdisciplinary fields such as Materials, Nanoscience, Chemical Engineering, Complex Systems and Biophysics.
- The work reported in the thesis must represent a significant scientific advance.
- If the thesis includes previously published material, permission to reproduce this must be gained from the respective copyright holder.
- They must have been examined and passed during the 12 months prior to nomination.
- Each thesis should include a foreword by the supervisor outlining the significance of its content.
- The theses should have a clearly defined structure including an introduction accessible to scientists not expert in that particular field.

Yilei Li

Probing the Response of Two-Dimensional Crystals by Optical Spectroscopy

Doctoral Thesis accepted
by Columbia University, New York, NY, USA

Springer

Yilei Li
Columbia University
New York, NY, USA

ISSN 2190-5053 ISSN 2190-5061 (electronic)
Springer Theses
ISBN 978-3-319-25374-9 ISBN 978-3-319-25376-3 (eBook)
DOI 10.1007/978-3-319-25376-3

Library of Congress Control Number: 2015953221

Springer Cham Heidelberg New York Dordrecht London

Printed on acid-free paper

Springer International Publishing AG Switzerland is part of Springer Science+Business Media (www.springer.com)

Supervisor's Foreword

Over the past decade, there have been remarkable advances in the synthesis and understanding of the properties of atomically thin 2D systems based on layered van der Waals (vdW) crystals. Starting with graphene, an ever-growing class of 2D materials has been identified, with many unusual and distinctive physical, electronic, optical, and chemical properties, as well as relevance for emerging technological applications. Moreover, the ability of combining these sheets into heterostructures with control at the monolayer level through mechanical assembly and growth techniques offers unprecedented possibilities for the development of new hybrid systems with properties tailored for fundamental science and applications.

Recent spectroscopic studies have begun to establish the basic properties of the electronic states and excitations in model 2D layers, including the semimetallic graphene layers, the semiconducting transition metal dichalcogenides (TMDCs), and the insulating hexagonal-BN layers. Dr. Yilei Li's thesis has made significant contributions to establishing the nature of electronic and vibrational excitations in these atomically thin layers. By applying linear and nonlinear optical spectroscopy, Dr. Li's thesis research has advanced our basic understanding of electronic properties and optical response of these atomically thin 2D layers, particularly for graphene and TMDC monolayers. Further, using Raman and infrared spectroscopy, Dr. Li has elucidated distinctive mechanisms that couple the electronic and vibration degrees of freedom in these 2D materials. Finally, Dr. Li's thesis presents the results of a magneto-optic study of monolayer $MoSe_2$ in the TMDC family. These investigations demonstrate the possibility of lifting the degeneracy between the K and K′ valleys in this material system, as well as the creation of a static valley polarization in doped layers. I would like to take this occasion to congratulate Dr. Yilei Li on the noteworthy research contributions made during his doctoral studies and to wish him further success in the coming years.

Menlo Park, CA, USA

Tony F. Heinz
SLAC National Accelerator Laboratory
Stanford University

Preface

Atomically thin 2D crystals form a distinct and growing class of new materials. The electromagnetic response of a 2D crystal provides direct access to its electronic properties. This thesis presents a series of experimental studies of the electromagnetic response of model 2D crystals as probed by optical spectroscopy. Our aim is to obtain understanding of their intrinsic linear and nonlinear response and the many-body interactions in these materials, as well as to explore the potential to use the 2D materials for sensing applications.

In the two studies of graphene, we either removed contaminations from the environment to reveal the intrinsic response or intentionally applied adsorbates to investigate how the electrons interact with the extrinsic molecules. In the first study, we obtained ultraclean graphene using hexagonal boron nitride as the substrate, which allowed us to probe using Raman spectroscopy the intrinsic electron–phonon and electron–electron interactions free from substrate induced sample inhomogeneity. In a second study, we demonstrate a strong near-field electromagnetic interaction of graphene plasmons with the vibrations of adsorbed molecules. Our results reveal the potential of graphene for molecular sensing.

In our investigations of the monolayer transition metal dichalcogenides, we performed measurements of the linear and the second-order nonlinear dielectric response. From the linear dielectric response, we demonstrate strong light-matter interactions even for a single layer of these materials. Several trends in the excitonic properties of this group of materials were obtained from the measured dielectric function. In the nonlinear optical study, we observed $>10^3$ enhancement of the second-harmonic signal from monolayers as compared to the bulk sample, a consequence of the breaking of the inversion symmetry present in the bulk. In addition to the results for monolayers, we describe the behavior of few-layer materials, where the symmetry properties change layer by layer. For monolayers (and samples of odd layer thickness with broken inversion symmetry), the strong and anisotropic second-harmonic response provides a simple optical probe of crystallographic orientation.

In the magneto-optic study of transition metal dichalcogenide monolayers, we demonstrate the induction of valley splitting and polarization by the application of an external magnetic field. The interaction of the valleys with the magnetic field reflects their nonzero magnetic moments, which are compared to theoretical models. We further clarify the electronic configuration of the charged excitons and important many-body corrections to the trion binding energy through the control of valley polarization achieved by the external magnetic field.

New York, NY Yilei Li

Acknowledgments

I am indebted to my advisor, Prof. Tony F. Heinz, for his guidance and support throughout the course of my thesis research. He has showed me by examples the scientific approach of breaking down complex problems into simpler subproblems that can be addressed more readily and directly. From him I learned the importance of developing clear physical insights in research. As an inspiring and knowledgeable advisor, he introduced to me the topics discussed in this thesis and opened up for me the field of optical spectroscopy of 2D crystals and beyond.

All of the members of Prof. Heinz's group have been generous in sharing their knowledge and their equipment with me. I should particularly mention Kin Fai Mak and Joshua Lui, who kindly introduced me to the group, lending me with numerous helps and set for me a high standard in graduate research. I have benefited much through a joint project with Yi Rao, who is especially motivating and supportive. More recently, Alexey Chernikov jointly pursued a number of studies with me. We had many discussions that are sparkled with interesting ideas, and we made good progress in our research as a result of these discussions. Working in the Heinz group has been a wonderful experience. I believe that the friendships built up here will continue beyond my graduate study.

It is an extremely valuable experience to have worked with many great collaborators outside our group: Inanc Meric, Cory Dean, Lei Wang, Arend van der Zande, Xian Zhang, Xiang Meng, and Leandro Malard at Columbia University; Hugen Yan, Damon Farmer, and Tony Low at IBM Research; Jonathan Ludwig, Prof. Zhiqiang Li, and Prof. Dmitry Smirnov at National High Magnetic Field Lab; Dr. Julia Stähler at Fritz Haber Institute; and Prof. Zonghai Hu at Peking University.

I would also like to take this opportunity to express my gratitude toward Dr. Phaedon Avouris, Prof. Richard Osgood, Prof. James Hone, Prof. Philip Kim, and Prof. Igor Aleiner who have shaped my graduate path with their guidance and wisdom and they will continue to be my best career references.

My heartfelt thanks are extended to my wife Jia Guo, who has contributed to this thesis through her delightful companionship and enduring cheer.

Contents

Chapter 1
Introduction and Background

1.1 Introduction

Low-dimensional systems, including 0-d (quantum dots), 1-d (nanotubes and nanowires), and 2-d (graphene, transition metal dichalcogenides, etc.) systems, exhibit many distinctive properties as compared to the bulk (three-dimensional) systems. The possibility of controlled production of many of these materials with good crystal quality and size suitable for experimental investigation has opened up many exciting avenues for the study of the fundamental physics in these low-dimensional systems. In general, the reduced dimensionality leads to a number of trends, including the enhancement of quantum confinement and the reduction in dielectric screening, both of which enhance interactions between charged particles in the material. On the other hand, the diverse material systems have already exhibited a wide variety of novel physical phenomenon, with potential for many more discoveries [1–4].

The measurement of a material's optical response can be categorized into two groups depending on the relative time scale of the temporal resolution of the measurement and the physical response of the material. In the first category, the measurement is slow compared to the material response. This is a measurement of the steady-state response. In the second category of time-resolved measurements, the temporal resolution of the measurement is fine enough to resolve the transient response of the sample under test. In this chapter, we focus on the first category because the main topics discussed in this thesis fall into this category. The time-resolved optical response of low-dimensional materials, while extremely interesting and important, will be largely left out for the purpose of achieving better focus. Several representative studies will be mentioned in the next section, where a review of the various optical spectroscopy methods is given.

Y. Li, *Probing the Response of Two-Dimensional Crystals by Optical Spectroscopy*, Springer Theses, DOI 10.1007/978-3-319-25376-3_1

1.2 Overview of Optical Spectroscopy Methods

In this section, we give a brief overview of the various optical techniques in the context of the study of low-dimensional materials. The optical techniques can be categorized by the frequencies of the excitations that they interact with (Fig. 1.1). In the terahertz (THz) frequency range, time-domain THz spectroscopy is a powerful tool, where the electric field (both amplitude and phase) of the THz signal is measured as a function of time [5]. Fourier transformation of the electric field in the time domain yields directly the THz spectrum in the frequency domain. A major area of interest using this technique is the study of the free carrier response, which arises from intraband electronic transitions [6]. Fourier transform IR (FTIR) spectrometry provides an alternative way to probe the terahertz response of a material [7]. There is one major limitation for FTIR measurement as compared to the time-domain THz measurement. FTIR spectroscopy provides only information on the light intensity, not the direct access to the electric field, as in the time-domain approach. As a result, to obtain the complex dielectric response, a Kramer–Kronig analysis is generally required.

In the far- to mid-IR range lie the lattice vibrational modes—phonons, the low-energy interband electronic transitions, and plasmonic resonances in 2D materials. FTIR spectroscopy probes these elementary excitations by either transmission or reflection measurement [8–12]. On the other hand, Raman spectroscopy probes these excitations by the inelastic Raman scattering process, where the difference between the incident photon energy and the emitted photon energy gives the energy of the phonon or other excited mode [13, 14].

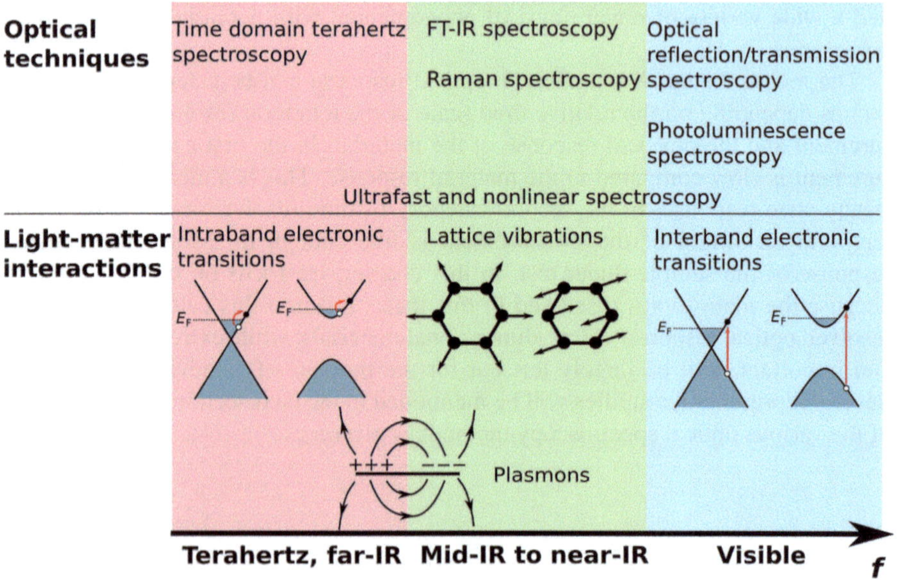

Fig. 1.1 Optical techniques and types of light-matter interactions in different frequency regimes

In the optical range, interband electronic transitions typically dominate the response and define the optical dielectric function of the material [15–17]. The optical dielectric response also provides valuable information on the strongly bound many-particle states, such as excitons and trions [18–22].

Ultrafast time-resolved pump-probe measurements using femtosecond lasers provide time resolution determined by the pulse duration. Depending on the wavelength of the pulse, it can be used to study the material response in different frequency regimes. In general, the pump pulse is much stronger than the probe pulse, so that the effect induced by the probe pulse itself can be neglected. The ultrafast optical techniques have been used widely to investigate the ultrafast phenomenon including the transient response of photoexcited carriers in graphene [23], the dynamics of hot phonons in graphene [24] and carbon nanotube [25] and the relaxation of photoexcited valley populations in monolayer transition metal dichalcogenides [26–31].

1.3 Optics of Graphene

Even though a major portion of this thesis is devoted to the study of monolayer transition metal dichalcogenide, we begin this introductory chapter with graphene, considering the relative simplicity of its electronic and optical properties, and the fact that graphene was the first material that drew the attention of the community to atomically thin materials [1].

Graphene, a single atomic layer unit of bulk graphite, consists of a honeycomb lattice of carbon atoms. The unit cell of the graphene lattice contains two inequivalent atomic sites, denoted as the A and B sites (Fig. 1.2a). The electronic structure of graphene was first calculated using a tight-binding model [32, 33]. The low energy electronic structure is described by a massless Dirac fermion Hamiltonian at the K and K′ points of the Brillouin zone (Fig. 1.2b):

$$\mathcal{H} = v_F \vec{\sigma} \cdot \vec{k}, \tag{1.1}$$

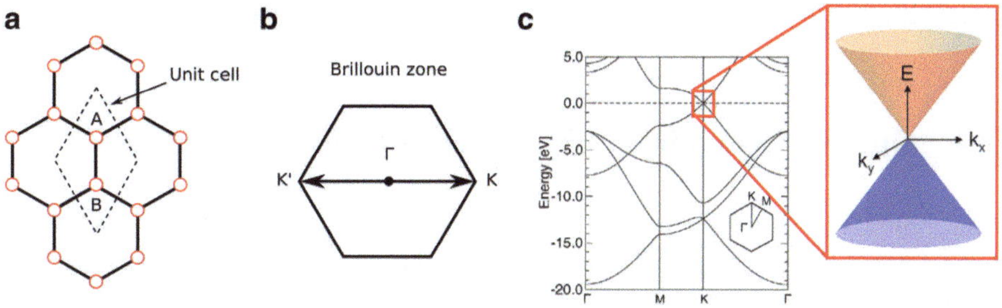

Fig. 1.2 (**a**) Honeycomb lattice of graphene. The unit cell contains two inequivalent (A and B) carbon atoms. (**b**) The Brillouin zone of graphene. The Dirac cones are located at the K and K′ points. (**c**) Electronic structure of graphene (adapted from Ref. [34])

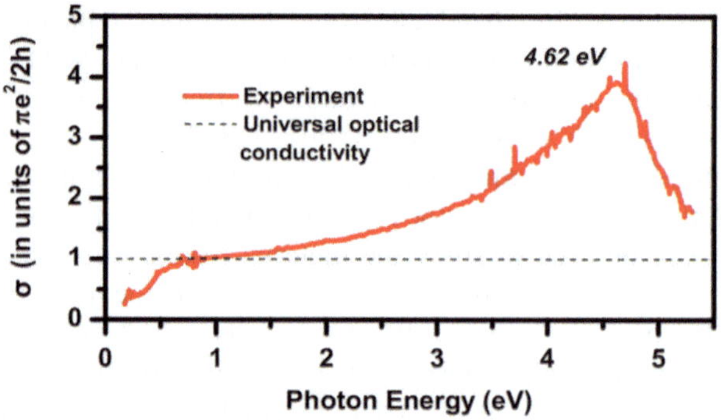

Fig. 1.3 Experimental optical conductivity (*solid line*) and the universal optical conductivity (*dashed line*) of monolayer graphene in the spectral range of 0.2–5.5 eV. The experimental peak energy is 4.62 eV. Note the deviation of the optical conductivity from the universal value at low energies is attributed to spontaneous doping [39]

where v_F is the Fermi velocity, $\vec{\sigma}$ denotes the Pauli matrices, and $\vec{k}$ is the crystal momentum. Intrinsic (undoped) graphene has its Fermi level located at the tip of the Dirac cone. We emphasis that the zero-gap and the linear dispersion is not model dependent and can be derived solely from symmetry considerations of the crystal structure [33]. For completeness, we also present the electronic structure of graphene in the whole Brillouin zone in Fig. 1.2c.

The Dirac Hamiltonian leads to the distinctive half-integer quantum hall effect and the linear dispersion results in quantum hall plateaus at energies proportional to $\mathrm{sgn}(N)\sqrt{|N|}$ where N is an integer and sgn (N) is the sign of N [35, 36]. Optically, the linear energy dispersion produces a constant interband contribution to the optical conductivity of $\sigma = \pi G_0 / 4$ (Fig. 1.3) [15, 37]. While the optical conductivity of graphene is close to this value in the photon energy range of 0.5–1.0 eV, it deviates significantly outside this frequency range. On the lower energy side, Pauli blocking results in a transparency below $2E_F$ [7, 9, 38], and the optical conductivity rises strongly in the far-infrared region from intraband contributions of the optical conductivity [7]. On the higher energy side, the deviation from linear dispersion leads to a significant increase and eventually a peak in the optical conductivity of graphene near 4.6 eV due to saddle point excitonic transitions [39].

1.4 Optics of Monolayer Transition Metal Dichalcogenides

The monolayer transition metal dichalcogenides MoS_2, $MoSe_2$, WS_2, and WSe_2 form a novel class of atomically thin materials that share many common structural, electronic, and optical properties. Very recently, monolayer $MoTe_2$ also joined the family [40]. In terms of crystal structure, they can be viewed as a single atomic layer section of the corresponding bulk materials—a hexagonal lattice of transition metal atoms sandwiched between two hexagonal lattices of chalcogen atoms in a

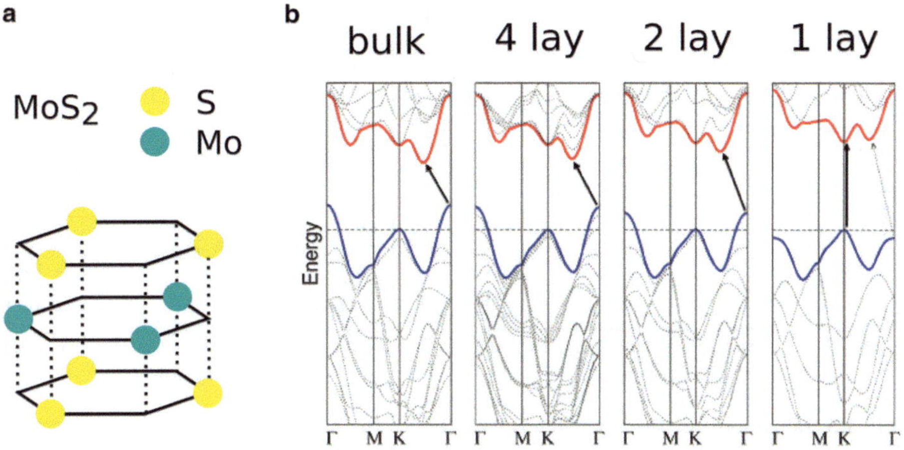

Fig. 1.4 (**a**) The lattice structure of monolayer MoS_2. (**b**) Electronic structure calculated by density functional theory for bulk, four layer, two layer, and single layer MoS_2. As the layer thickness goes down, the indirect gap opens up and single layer MoS_2 becomes a direct gap semiconductor [43]

trigonal prismatic arrangement (Fig. 1.4a)] [41]. Similar to the bulk, the interband transitions at the K and K′ points of the Brillouin zone of monolayer TMDCs, where the valence band is split by spin-orbit coupling, give rise to the A and B peaks in the optical absorption spectra [42]. On the other hand, the monolayers exhibit a number of important physical properties that are very distinct from the corresponding bulk materials. Monolayer MoS_2 is a direct-gap semiconductor, in contrast to the indirect-gap bulk MoS_2. This indirect to direct band gap transition was first observed through the shift of the indirect PL feature with layer number and a dramatic increase in the photoluminescence efficiency in the monolayer compared to the bilayer and the bulk [42, 43]. Figure 1.4b shows the electronic structures of monolayer, bilayer, four layer, and bulk MoS_2 calculated by density functional theory, confirming the direct-gap character of the monolayer.

In a two-dimensional system with parabolic energy dispersion for electrons and holes, the joint density of states near the band gap is a step function [44]. However, as shown in Fig. 1.5a, the measured absorption spectra of the monolayer TMDCs exhibit peaked features at the band edge, implying strong excitonic effects in these transitions. In fact, Rydberg series of the excitonic transitions is identified by reflectance contrast spectroscopy, as shown in Fig. 1.5b [18, 19]. The energy spacing between the Rydberg levels is found to be non-hydrogenic and a large exciton binding energy of 300–400 meV is determined in the tungsten dichalcogenide monolayers [18, 19]. At finite doping, a second peak, identified as the trion peak, appears on the lower energy side of the neutral exciton peak, the strength of which increases with increasing doping [20, 21]. The energy difference between the two peaks grows linearly with the Fermi level, a signature of charged excitons that stems from the doping-induced change in the many-body binding energy of the trion.

In monolayer transition metal dichalcogenides, the K and K′ valleys correspond to distinct electronic states with opposite momenta. The two valleys were predicted

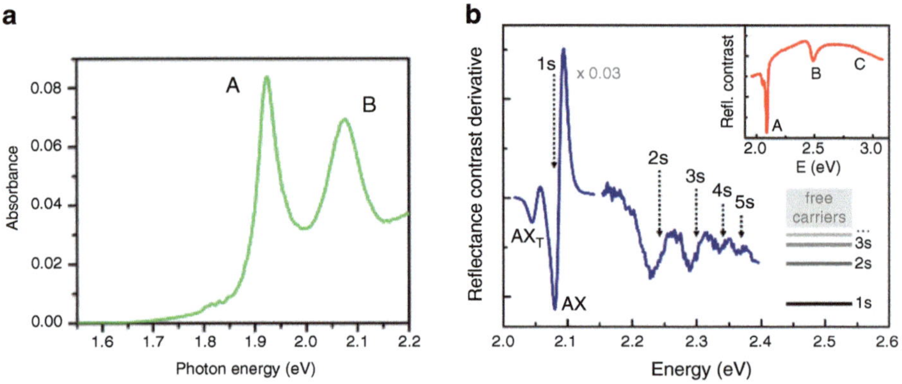

Fig. 1.5 (**a**) Absorbance spectrum of monolayer MoS_2. The A and B peaks labeled in the spectrum correspond to the spin-split interband transitions from the K and K′ points [20]. (**b**) First derivative of the reflectance contrast spectrum of monolayer WS_2. The 1 s excitons (neutral exciton and trion) and the higher excited exciton states (Rydberg series) are labeled. The *inset* shows the original reflectance contrast spectrum [18]

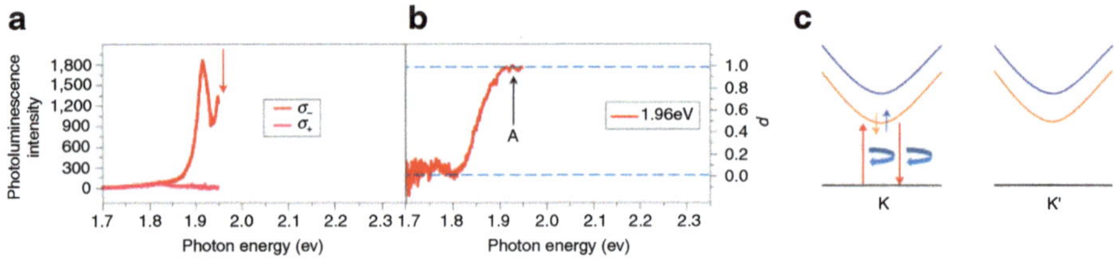

Fig. 1.6 (**a**) Left circularly polarized (σ_-) and right circularly polarized (σ_+) photoluminescence spectra for σ_- excitation at 1.96 eV (633 nm). (**b**) Photoluminescence helicity as a function of the photon energy. (**c**) Schematic representation of the valley-selective optical absorption and emission processes [46]

theoretically to be independently addressable by optical helicity [45], as was then demonstrated experimentally by circular-polarization resolved photoluminescence spectroscopy [3, 46, 47] (Fig. 1.6). By pumping one of the two valleys using circular polarized light, unequal transient valley populations can be created.

References

1. Geim, A.K., Novoselov, K.S.: The rise of graphene. Nat. Mater. **6**, 183–191 (2007)
2. Butler, S.Z., Hollen, S.M., Cao, L., Cui, Y., Gupta, J.A., Gutiérrez, H.R., Heinz, T.F., Hong, S.S., Huang, J., Ismach, A.F., Johnston-Halperin, E., Kuno, M., Plashnitsa, V.V., Robinson, R.D., Ruoff, R.S., Salahuddin, S., Shan, J., Shi, L., Spencer, M.G., Terrones, M., Windl, W., Goldberger, J.E.: Progress, challenges, and opportunities in two-dimensional materials beyond graphene. ACS Nano **7**, 2898–2926 (2013)

3. Cao, T., Wang, G., Han, W., Ye, H., Zhu, C., Shi, J., Niu, Q., Tan, P., Wang, E., Liu, B., Feng, J.: Valley-selective circular dichroism of monolayer molybdenum disulphide. Nat. Commun. **3**, 887 (2012)
4. Xu, X., Yao, W., Xiao, D., Heinz, T.F.: Spin and pseudospins in layered transition metal dichalcogenides. Nat. Phys. **10**, 343–350 (2014)
5. Grischkowsky, D., Keiding, S.R., van Exter, M., Fattinger, C.: Far-infrared time-domain spectroscopy with terahertz beams of dielectrics and semiconductors. J. Opt. Soc. Am. B **7**, 2006–2015 (1990)
6. Jnawali, G., Rao, Y., Yan, H., Heinz, T.F.: Observation of a transient decrease in terahertz conductivity of single-layer graphene induced by ultrafast optical excitation. Nano Lett. **13**, 524–530 (2013)
7. Horng, J., Chen, C.-F., Geng, B., Girit, C., Zhang, Y., Hao, Z., Bechtel, H.A., Martin, M., Zettl, A., Crommie, M.F., Shen, Y.R., Wang, F.: Drude conductivity of Dirac fermions in graphene. Phys. Rev. B **83**, 165113 (2011)
8. Mak, K.F., Sfeir, M.Y., Misewich, J.A., Heinz, T.F.: The evolution of electronic structure in few-layer graphene revealed by optical spectroscopy. Proc. Natl. Acad. Sci. **107**, 14999–15004 (2010)
9. Wang, F., Zhang, Y., Tian, C., Girit, C., Zettl, A., Crommie, M., Shen, Y.R.: Gate-variable optical transitions in graphene. Science **320**, 206–209 (2008)
10. Ju, L., Geng, B., Horng, J., Girit, C., Martin, M., Hao, Z., Bechtel, H.A., Liang, X., Zettl, A., Shen, Y.R., Wang, F.: Graphene plasmonics for tunable terahertz metamaterials. Nat. Nano **6**, 630–634 (2011)
11. Zhang, Y., Tang, T.-T., Girit, C., Hao, Z., Martin, M.C., Zettl, A., Crommie, M.F., Shen, Y.R., Wang, F.: Direct observation of a widely tunable bandgap in bilayer graphene. Nature **459**, 820–823 (2009)
12. Tang, T.-T., Zhang, Y., Park, C.-H., Geng, B., Girit, C., Hao, Z., Martin, M.C., Zettl, A., Crommie, M.F., Louie, S.G., Shen, Y.R., Wang, F.: A tunable phonon-exciton Fano system in bilayer graphene. Nat. Nano **5**, 32–36 (2010)
13. Ferrari, A.C., Meyer, J.C., Scardaci, V., Casiraghi, C., Lazzeri, M., Mauri, F., Piscanec, S., Jiang, D., Novoselov, K.S., Roth, S., Geim, A.K.: Raman spectrum of graphene and graphene layers. Phys. Rev. Lett. **97**, 187401 (2006)
14. Malard, L.M., Pimenta, M.A., Dresselhaus, G., Dresselhaus, M.S.: Raman spectroscopy in graphene. Phys. Rep. **473**, 51–87 (2009)
15. Mak, K.F., Sfeir, M.Y., Wu, Y., Lui, C.H., Misewich, J.A., Heinz, T.F.: Measurement of the optical conductivity of graphene. Phys. Rev. Lett. **101**, 196405 (2008)
16. Nair, R.R., Blake, P., Grigorenko, A.N., Novoselov, K.S., Booth, T.J., Stauber, T., Peres, N.M.R., Geim, A.K.: Fine structure constant defines visual transparency of graphene. Science **320**, 1308 (2008)
17. Li, Y., Chernikov, A., Zhang, X., Rigosi, A., Hill, H., van der Zande, A. M., Chenet, D.A., Shih, E.-M., Hone, J., Heinz, T.F.: Measurement of the optical dielectric function of monolayer transition-metal dichalcogenides: MoS_2, $MoSe_2$, WS_2, and WSe_2. Phys. Rev. B **90**, 205422 (2014)
18. Chernikov, A., Berkelbach, T.C., Hill, H.M., Rigosi, A., Li, Y., Aslan, O.B., Reichman, D.R., Hybertsen, M.S., Heinz, T.F.: Exciton binding energy and nonhydrogenic rydberg series in monolayer WS2. Phys. Rev. Lett. **113**, 076802 (2014)
19. He, K., Kumar, N., Zhao, L., Wang, Z., Mak, K.F., Zhao, H., Shan, J.: Tightly bound excitons in monolayer WSe2. Phys. Rev. Lett. **113**, 026803 (2014)
20. Mak, K.F., He, K., Lee, C., Lee, G.H., Hone, J., Heinz, T.F., Shan, J.: Tightly bound trions in monolayer MoS_2. Nat. Mater. **12**, 207–211 (2013)
21. Ross, J.S., Wu, S., Yu, H., Ghimire, N.J., Jones, A.M., Aivazian, G., Yan, J., Mandrus, D.G., Xiao, D., Yao, W., Xu, X.: Electrical control of neutral and charged excitons in a monolayer semiconductor. Nat. Commun. **4**, 1474 (2013)
22. Ye, Z., Cao, T., O'Brien, K., Zhu, H., Yin, X., Wang, Y., Louie, S.G., Zhang, X.: Probing excitonic dark states in single-layer tungsten disulphide. Nature **513**, 214–218 (2014)
23. Leandro, M.M., Kin Fai, M., Neto, A.H.C., Peres, N.M.R., Tony, F.H.: Observation of intra- and inter-band transitions in the transient optical response of graphene. New J. Phys. **15**, 015009 (2013)

24. Wu, S., Liu, W.-T., Liang, X., Schuck, P.J., Wang, F., Shen, Y.R., Salmeron, M.: Hot phonon dynamics in graphene. Nano Lett. **12**, 5495–5499 (2012)
25. Song, D., Wang, F., Dukovic, G., Zheng, M., Semke, E.D., Brus, L.E., Heinz, T.F.: Direct measurement of the lifetime of optical phonons in single-walled carbon nanotubes. Phys. Rev. Lett. **100**, 225503 (2008)
26. Wang, Q., Ge, S., Li, X., Qiu, J., Ji, Y., Feng, J., Sun, D.: Valley carrier dynamics in monolayer molybdenum disulfide from helicity-resolved ultrafast pump–probe spectroscopy. ACS Nano **7**, 11087–11093 (2013)
27. Mai, C., Barrette, A., Yu, Y., Semenov, Y.G., Kim, K.W., Cao, L., Gundogdu, K.: Many-body effects in valleytronics: direct measurement of valley lifetimes in single-layer MoS_2. Nano Lett. **14**, 202–206 (2014)
28. Kumar, N., Cui, Q., Ceballos, F., He, D., Wang, Y., Zhao, H.: Exciton-exciton annihilation in $MoSe_2$ monolayers. Phys. Rev. B **89**, 125427 (2014)
29. Sun, D., Rao, Y., Reider, G.A., Chen, G., You, Y., Brézin, L., Harutyunyan, A.R., Heinz, T.F.: Observation of rapid exciton–exciton annihilation in monolayer molybdenum disulfide. Nano Lett. **14**, 5625–5629 (2014)
30. Cui, Q., Ceballos, F., Kumar, N., Zhao, H.: Transient absorption microscopy of monolayer and bulk WSe2. ACS Nano **8**, 2970–2976 (2014)
31. Kim, J., Hong, X., Jin, C., Shi, S.-F., Chang, C.-Y.S., Chiu, M.-H., Li, L.-J., Wang, F.: Ultrafast generation of pseudo-magnetic field for valley excitons in WSe2 monolayers. Science **346**, 1205–1208 (2014)
32. Wallace, P.R., Band, T.: Theory of graphite. Phys. Rev. **71**, 622–634 (1947)
33. Slonczewski, J.C., Weiss, P.R.: Band structure of graphite. Phys. Rev. **109**, 272–279 (1958)
34. Latil, S., Henrard, L.: Charge carriers in few-layer graphene films. Phys. Rev. Lett. **97**, 036803 (2006)
35. Zhang, Y., Tan, Y.-W., Stormer, H.L., Kim, P.: Experimental observation of the quantum Hall effect and Berry's phase in graphene. Nature **438**, 201–204 (2005)
36. Novoselov, K.S., Geim, A.K., Morozov, S.V., Jiang, D., Katsnelson, M.I., Grigorieva, I.V., Dubonos, S.V., Firsov, A.A.: Two-dimensional gas of massless Dirac fermions in graphene. Nature **438**, 197–200 (2005)
37. Kuzmenko, A.B., van Heumen, E., Carbone, F., van der Marel, D.: Universal Optical Conductance of Graphite. Phys. Rev. Lett. **100**, 117401 (2008)
38. Li, Z.Q., Henriksen, E.A., Jiang, Z., Hao, Z., Martin, M.C., Kim, P., Stormer, H.L., Basov, D.N.: Dirac charge dynamics in graphene by infrared spectroscopy. Nat. Phys. **4**, 532–535 (2008)
39. Mak, K.F., Shan, J., Heinz, T.F.: Seeing many-body effects in single- and few-layer graphene: observation of two-dimensional saddle-point excitons. Phys. Rev. Lett. **106**, 046401 (2011)
40. Ruppert, C., Aslan, O.B., Heinz, T.F.: Optical properties and band gap of single- and few-layer MoTe2 crystals. Nano Lett **14**, 6231–6236 (2014)
41. Wilson, J.A., Yoffe, A.D.: The transition metal dichalcogenides discussion and interpretation of the observed optical, electrical and structural properties. Adv. Phys. **18**, 193–335 (1969)
42. Mak, K.F., Lee, C., Hone, J., Shan, J., Heinz, T.F.: Atomically thin MoS_2: a new direct-gap semiconductor. Phys. Rev. Lett. **105**, 136805 (2010)
43. Splendiani, A., Sun, L., Zhang, Y., Li, T., Kim, J., Chim, C.-Y., Galli, G., Wang, F.: Emerging photoluminescence in monolayer MoS_2. Nano Lett **10**, 1271–1275 (2010)
44. Klingshirn, C.F.: Semiconductor optics. Springer, Berlin (2007)
45. Xiao, D., Liu, G.-B., Feng, W., Xu, X., Yao, W.: Coupled spin and valley physics in monolayers of MoS_2 and other group-VI dichalcogenides. Phys. Rev. Lett. **108**, 196802 (2012)
46. Mak, K.F., He, K., Shan, J., Heinz, T.F.: Control of valley polarization in monolayer MoS_2 by optical helicity. Nat. Nano **7**, 494–498 (2012)
47. Zeng, H., Dai, J., Yao, W., Xiao, D., Cui, X.: Valley polarization in MoS_2 monolayers by optical pumping. Nat. Nano **7**, 490–493 (2012)

Chapter 2
Intrinsic Doping Dependence of Raman 2D Mode in Graphene: Signatures of Electron–Electron Interaction

2.1 Introduction

Raman spectroscopy has been one of the most useful experimental probes for both fundamental study and characterization of graphene thin layers [1, 2]. Raman 2D mode in graphene is originated from a coherent process that involves the scattering of the quasiparticles twice by near K-point phonons [3, 4]. Resonant with the Dirac cones, the 2D mode is sensitive to perturbations in the electronic states, and this property has made it a versatile diagnostic tool for characterizing layer number and stacking order of few-layer graphene [2, 5–7]. Apart from the above applications, the doping dependence of the Raman 2D mode is of particular interest as it provides a monitor for the charge density in graphene [8]. Here we reinvestigate this problem, but using ultraclean graphene exfoliated on hexagonal boron nitride (h-BN), which provides excellent charge homogeneity [9–11]. Contrary to the weak doping dependence of Raman 2D mode below a net charge density of 2×10^{12} cm^{-2} in [8], we observed a significant blue shift of more than 4 cm^{-1} when the sample is gated over the same density range. Because of the strong doping dependence of the intrinsic shift, improved charge homogeneity is essential for proper measurement of the properties of this feature at low doping levels. The observed symmetric blue shift of 2D mode is shown to be a direct consequence of the density-dependent renormalization of electronic band structure by electron–electron (*e–e*) interactions [12–20]. Interestingly, direct renormalization of the phonon frequency with doping is found to be relatively weak for this range of doping and to imply a shift in frequency opposite to what has been observed experimentally [21].

In addition to the peak position shift, the 2D mode broadens by 24 cm^{-1} per eV doping, and its intensity decreases by 20 % at our highest doping level of 2×10^{12} cm^{-2}. The observed broadening and intensity reduction of 2D mode are also signatures of *e–e* interactions in graphene: enhanced *e–e* scattering with doping [19] reduces the lifetime of photo-excited electron–hole (*e–h*) pairs, which makes the resonant condition less stringent and thereby increases the width of the feature and

Y. Li, *Probing the Response of Two-Dimensional Crystals by Optical Spectroscopy*, Springer Theses, DOI 10.1007/978-3-319-25376-3_2

decreases its strength. By identifying *e–e* interactions as the predominant cause of the doping dependence of graphene 2D mode, we demonstrate the double-resonant Raman mode as a sensitive probe of change in electronic self-energy from *e–e* interactions. Interestingly, renormalization of the phonon frequency with doping is found to be relatively weak for this range of doping and to imply a shift in frequency opposite to what has been observed experimentally [22]. In contrast to previous complementary experimental approaches for the study of electronic structure renormalization by *e–e* interactions near the Fermi level [12, 14, 15], we are able to probe here the properties of quasiparticles more than 1 eV above the Fermi level.

2.2 Experimental Details for Raman Measurement of Ultraclean Graphene on h-BN

Our Raman measurements were performed on graphene field-effect transistors (FETs) fabricated using h-BN as the gate dielectric [23]. The device structure is illustrated in Fig. 2.1a. A piece of exfoliated graphene of size 2 μm × 5 μm was transferred onto an h-BN layer of 16 nm thickness. Three metal electrodes were used in our FET device, with one of them placed under the h-BN layer to electrostatically gate graphene and the other two placed on top of graphene as source and drain electrodes. Figure 2.1b shows an optical image of our device. Its electrical characteristics were examined through measurement of the device resistance at low bias $(V_{sd} = 0.1\text{V})$ as a function of gate voltage V_{gs} (Fig. 2.1c). Electron/hole mobilities

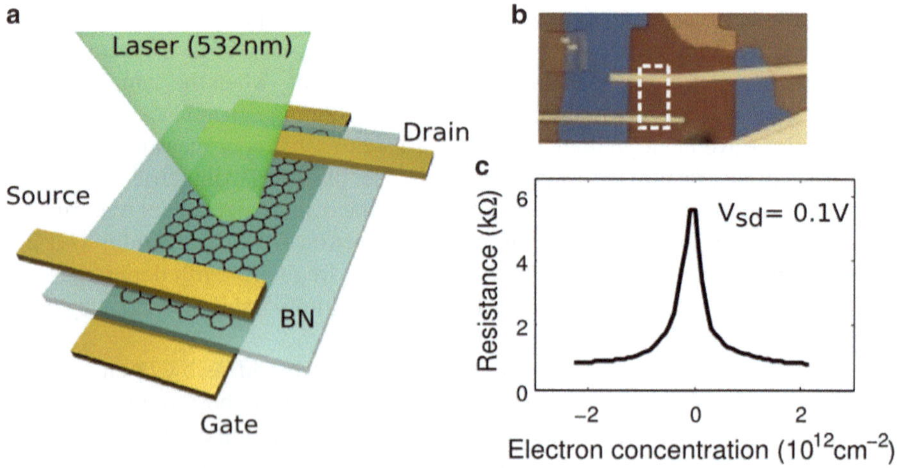

Fig. 2.1 (**a**) Schematic representation of the graphene FET device using an h-BN gate dielectric under laser illumination. (**b**) Optical image of the graphene FET device. The area occupied by the graphene layer is indicated by the *dashed lines*. (**c**) Measured I–V characteristics of our graphene FET device. The carrier mobility near the charge neutrality point exceeds 1.7×10^4 cm^2 V^{-1} s^{-1}

exceeding 1.7×10^4 cm^2 V^{-1} s^{-1} were extracted from the slope of doping dependence of the device resistance near charge neutrality at room temperature. From the width of the same curve, we obtain an upper bound for the charge inhomogeneity in our sample of 1.5×10^{11} cm^{-2}. The Raman spectra of the gated graphene were obtained using a commercial Raman microscope (Renishaw). The excitation source was a 532 nm laser, focused to a spot size of ~0.5 μm on the sample and having power of 0.2 mW. The spectral resolution for the Raman measurements was <1 cm^{-1}.

2.3 Intrinsic Doping Dependence of the Raman G Mode: Signatures of *e–p* Interactions

The dominant Raman features in high-quality graphene are the G and 2D modes [1, 2]. Although our primary focus will be the 2D mode, we briefly discuss the behavior observed for the G mode as a function of doping level. The G peak is a first-order Raman process (Fig. 2.2a) associated with the emission of a zone-center optical phonon. In several recent experimental studies, a narrowing and blue shift of the G mode (Fig. 2.2c) were observed when the Fermi level of graphene is tuned away from the Dirac point [8, 24–30]. These phenomena are understood to arise from the Pauli blocking of electron–phonon (*e–p*) coupling when doping prevents decay of an optical

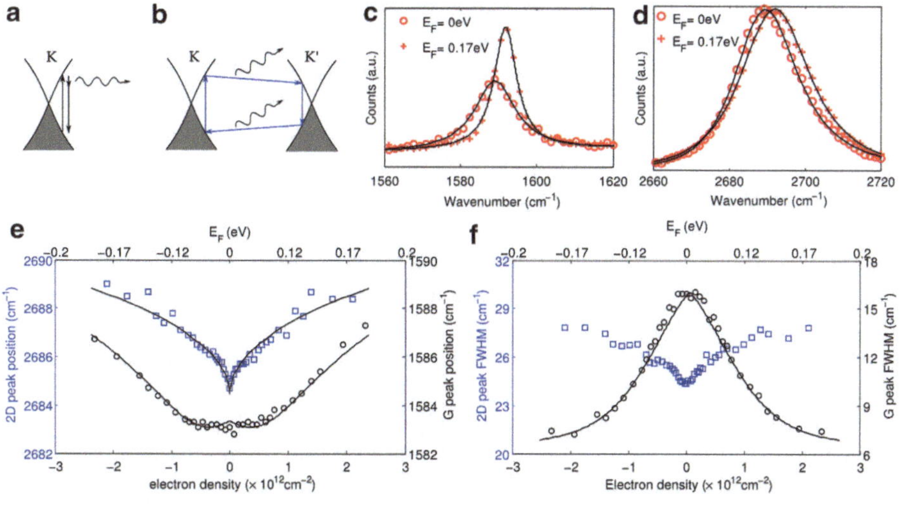

Fig. 2.2 (**a, b**) Illustration of Raman scattering processes corresponding to Raman G and 2D modes. (**c, d**) Measured G and 2D spectra (symbols) at charge neutrality point ($E_F = 0$) and at our highest doping level ($E_F = 0.17$ eV). The *black lines* are theoretical fits as described in the text. The 2D peak heights are normalized for clearer comparison. (**e**) Measured frequencies of the Raman 2D (*blue squares*) and G (*black circles*) modes versus charge density in the graphene layer. The theoretical results are shown as *black solid lines*. (**f**) Measured widths of the Raman 2D (*blue squares*) and G peak (*black circles*) modes as functions of charge density

phonon into an electron–hole pair [31–33]. However, appreciable differences are generally observable between experiments and theoretical calculations, likely due to experimental uncertainties. With the ultraclean graphene sample on BN, our measured G mode position and FWHM doping dependencies at room temperature conform to the perturbative calculations almost perfectly (Fig. 2.2e, f), reconfirming superior device quality and very good experimental control of doping level in graphene.

2.4 Intrinsic Doping Dependence of the Raman 2D Mode: Signatures of *e–e* Interactions

The Raman 2D mode, unlike G mode, is originated from a coherent double resonant process [3, 4], as illustrated in Fig. 2.2b. It involves the excitation of an *e–h* pair near K point by absorbing a photon, the inter-cone scattering of the charged particle twice by near K point phonon and the emission of a Stokes photon by *e–h* recombination near K′ point. One important consequence of the double resonant process is that phonons involved in the Raman process are selected by the resonant condition, and as a result, a change in the resonance condition will lead to a change in the selected phonons in momentum space. Comparing to the G mode case, the doping dependence of the 2D mode is much less understood. A recent study of graphene on SiO_2 [22] shows a red shift of ~16 cm^{-1}/eV and a broadening of ~12 cm^{-1}/eV in 2D mode with doping. The shift of the 2D mode observed in [22] is in the opposite direction of the 2D mode shift we observed for graphene on h-BN and the peak broadening observed in [22] is smaller. These differences may reflect the role of different substrates used in the experiments. In graphene with low charge inhomogeneity, slight doping by a few hundred meV cannot block the resonant *e–p* scattering processes far away from the Dirac point. Perturbative calculation shows that the position shift of the Raman 2D mode with doping due to modified *e–p* coupling is smaller than 0.2 cm^{-1} at the doping level of $2\times10^{-12}\,cm^{-2}$ [21].

As the line-shape of the Raman 2D mode for intrinsic graphene is slightly asymmetric [34–37], we extract the doping-dependent position and full-width at half maximum (FWHM) by fitting the spectra to a phenomenological asymmetric function (Fig. 2.2d). We found the 2D mode blue shifts for either electron or hole doping (Fig. 2.2e). Moreover, the shifts on the electron and hole sides are essentially symmetrical, with an absolute peak shift per eV doping of 25 cm^{-1}/eV.

Since the renormalization of phonon energy is expected to be too small to account for the observed blue shift of the 2D peak, we turn our attentions to an important effect for the resonant Raman process, the renormalization of electronic bands. Prevailing experimental evidences have shown that Coulomb interactions among the charged carriers renormalize the graphene's electronic structure and enhance the Fermi velocity [12–20]. Doping reduces the absolute value of electron energies and the resonance condition pushes the vertical electronic transition away from the K and K′ points, thus setting the resonant phonons further away from the K point (Fig. 2.3). Because of the anomalous phonon energy dispersion near the K point

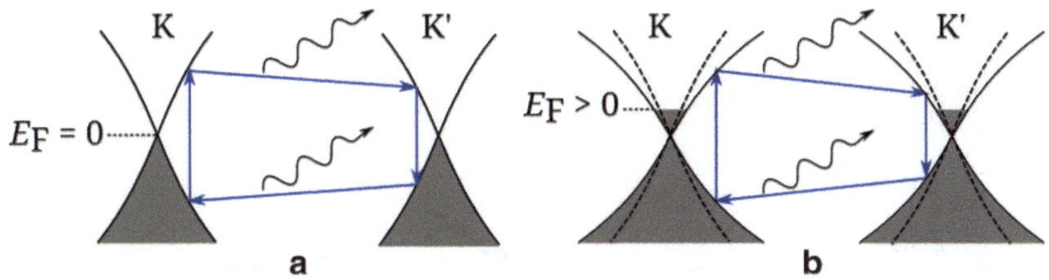

Fig. 2.3 *e–e* interactions responsible for the symmetric blue shift of Raman 2D mode in the doping range below 2×10^{12} cm^{-2}. (**a**) Schematic in momentum space of the resonant Raman 2D scattering process in undoped graphene showing the electronic structure of graphene near the K and K′ points. Two near K point phonons are emitted in this process. The *blue arrows* describe possible scattering pathways of charge carriers. (**b**) At finite doping, the electronic bands of graphene (*solid lines*) become flatter compared to the undoped case (*dashed lines*). The resonant electronic processes for the Raman 2D mode are then pushed further away from the K and K′ points and involve phonons of decreased momentum

[31, 38, 39], the 2D Raman scattering process involves phonons of higher energies, increasing the 2D mode Raman shift. Far above the Fermi surface, the self-energy correction due to *e–e* interaction has been calculated theoretically [14, 20, 40–44]

$$\Re\Sigma_{e-e}(k) = \frac{\alpha}{4}\hbar\upsilon_F^0 |k-k_F| \ln\left(\left|\frac{k_c}{k-k_F}\right|\right), \tag{2.1}$$

where α is the coupling constant in graphene and υ_F^0 is the Fermi velocity of the bare electrons. k is the momentum of the quasiparticles, k_F is the Fermi vector, and k_c is the ultraviolet cutoff momentum taken to be $1.703\,\text{Å}^{-1}$. Note that we have approximated the electron gas to be degenerate and neglected finite temperature effect since the scale of the ambient temperature is comparable to our doping step size. To understand from (2.1) how does the 2D mode shift depend on other parameters, we approximate the logarithmic term to be 1, which is appropriate in our case where k is comparable to k_c. We then apply the double resonance condition and obtain $\frac{d\hbar\omega_0}{dE_F} < \alpha\frac{\beta_K}{v_F}$ where β_K is the slope of graphene's phonon band structure near the K point and v_F is the renormalized Fermi velocity. Quantitatively, we fit our experimental data to (2.1) using $\upsilon_F^0 = 10^6\,\text{m/s}$ and $\beta_K = 940\,\text{cm}^{-1}$ (determined from the 2D mode dispersion with excitation energy). α is treated as an adjustable fitting parameter. A very good fitting to our experimental data is achieved with $\alpha = 0.60$ (Fig. 2.2e). This value of α agrees with the values of α for graphene on h-BN obtained from other experiments [18, 45]. At much higher doping level, however, the modification of phonon energy by the change in atomic bond lengths becomes increasingly significant and the 2D mode shifts to the red side [8].

We can check the consistency of the band renormalization picture with the well-known dispersion of 2D mode with excitation photon energy [46, 47] by comparing the peak shift in 2D mode with doping at different excitation energies (Fig. 2.4a).

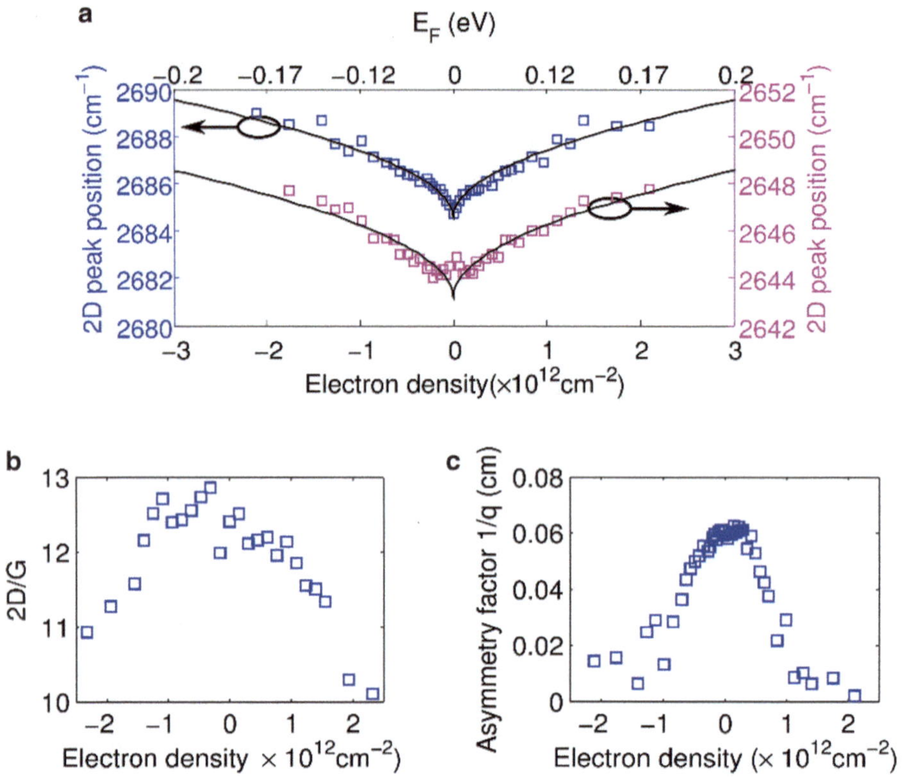

Fig. 2.4 (**a**) 2D mode position versus doping excited by photons at 2.33 eV (*blue squares*) and 1.96 eV (*magenta squares*). The *solid lines* are theoretical models taking into account the effects of electron energy renormalization due to *e–e* interactions as discussed in the text. (**b, c**) Measured 2D/G intensity ratio and the asymmetry factor 1/q as a function of electron density

We fit our experimental results with 2.33 eV excitation and 1.96 eV excitation simultaneously using the double resonance picture. We used two fitting parameters, the phonon energy exactly at K point E_K and α. The 2D peak shift by doping and the 2D peak dispersion with excitation energies can be understood simultaneously using $E_K = 1229\text{cm}^{-1}$ and $\alpha = 0.60$.

Our measured shift in the 2D mode position is in apparent contradiction with the experimental studies using electrochemical gating [8], where the 2D mode position has variation smaller than 1 cm^{-1} when the absolute charge density is below 2×10^{12} cm^2. We can reconcile these two pictures by noting that with electrochemical gating, the higher charge inhomogeneity is likely to smear out the sharp non-analytical feature near the charge neutrality point. In addition, the extra layer of electrolyte can enhance the screening of *e–e* interactions, thus reducing the interaction effect in the Raman spectra. At higher doping levels, however, even in the presence of a relatively large amount of charge inhomogeneity, the experimental observed variation in Raman 2D peak position shift with doping should represent closely the intrinsic doping dependence [8] due to the absence of any abrupt non-monotonic change in the 2D position. On a practical aspect, our results show that it is not appropriate to use 2D mode position to predict the sign of doping in graphene in the doping region below 3×10^{12} cm^{-2}. A comparison of the G and 2D peak positions doping dependencies clearly indicates that the non-analytic shift of 2D mode is more susceptible to inhomogeneity in doping concentration than the relatively flat G mode doping dependence within 10^{12} cm^{-2} from the Dirac point (Fig. 2.2e).

In addition to the Raman 2D mode peak shift, the peak FWHM is broadened by ~4 cm^{-1} when the Fermi level is shifted from the Dirac point by 0.2 eV (Fig. 2.2f). The width of the Raman 2D peak is induced mainly by the interference of all the possible Raman scattering pathways satisfying the double resonance condition instead of the intrinsic phonon lifetime. It is natural then that the finite electronic line width contributes to the total line width of the Raman 2D mode. This is because at any fixed momentum, there is a distribution of electronic states with different energies, loosing the constraint of double resonance condition. Charge doping increases the phase space for the *e–e* scattering processes, and reduces *e–h* pair lifetime. Thus it is expected that doping will broaden the Raman 2D peak. In a similar way as we estimate the 2D peak shift, we can estimate a lower bound for the broadening of electronic states from the peak broadening: $\Delta_{\gamma_e} \gtrsim \frac{1}{4}\Delta_\gamma \upsilon_F / \beta = 10\text{meV}$, where the factor $\frac{1}{4}$ accounts for the initial and final states of the two phonon emission events. The lower bound of electronic broadening is about 10 % of the experimental measured quasiparticle lifetime of ~0.1 eV, near 1 eV above or below the Fermi level [20, 48]. This agrees with the 20 % decrease of 2D/G intensity ratio with doping (Fig. 2.4b), as it is predicted theoretically that the 2D/G intensity ratio is proportional to inverse square of the quasiparticle lifetime [19].

Slight asymmetry of the Raman 2D mode was observed near the charge neutrality point [49] for our graphene sample on BN substrate, similar to that observed in suspended graphene with ultralow net charge density [34, 35, 37]. Theoretical calculations show that the interference among all the Raman scattering pathways could result in qualitatively similar asymmetry [4, 36]. Interestingly, as the Fermi level is moved away from the charge neutrality point, we observed that the 2D line-shape becomes increasingly symmetrical (Fig. 2.4c). We note that excitonic effect may give further corrections to the simple band picture. It can potentially give corrections to the position, width, and even the line shape of the 2D mode. This is, however, beyond the scope of discussion in this work.

2.5 Conclusion

In summary, we measured the doping dependence of the Raman G and 2D modes in ultraclean graphene on h-BN substrate. The G mode doping dependence shows similar features as in previous studies, but the better agreement with theory confirms the outstanding sample quality. The 2D mode shifts symmetrically by 25 cm^{-1} per eV doping to higher energy below a doping level of 2×10^{12} cm^{-2}. We attribute this symmetric blue shift to the change in double resonance condition through the renormalization of electronic band structure by *e–e* interactions. The width of the graphene 2D mode increases with doping by 24 cm^{-1} per eV doping while the 2D/G intensity ratio decreases with doping by 20 % at our highest doping level. These two trends of the Raman 2D mode with doping are shown to be natural consequences of the reduction of photo-excited carrier lifetime.

References

1. Malard, L.M., Pimenta, M.A., Dresselhaus, G., Dresselhaus, M.S.: Raman spectroscopy in graphene. Phys. Rep. **473**, 51–87 (2009)
2. Ferrari, A.C., Meyer, J.C., Scardaci, V., Casiraghi, C., Lazzeri, M., Mauri, F., Piscanec, S., Jiang, D., Novoselov, K.S., Roth, S., Geim, A.K.: Raman spectrum of graphene and graphene layers. Phys. Rev. Lett. **97**, 187401 (2006)
3. Thomsen, C., Reich, S.: Double resonant Raman scattering in graphite. Phys. Rev. Lett. **85**, 5214–5217 (2000)
4. Maultzsch, J., Reich, S., Thomsen, C.: Double-resonant Raman scattering in graphite: interference effects, selection rules, and phonon dispersion. Phys. Rev. B **70**, 155403 (2004)
5. Gupta, A., Chen, G., Joshi, P., Tadigadapa, S., Eklund, P.C.: Raman scattering from high-frequency phonons in supported n-graphene layer films. Nano Lett. **6**, 2667–2673 (2006)
6. Lui, C.H., Li, Z., Chen, Z., Klimov, P.V., Brus, L.E., Heinz, T.F.: Imaging stacking order in few-layer graphene. Nano Lett. **11**, 164–169 (2010)
7. Cong, C., Yu, T., Sato, K., Shang, J., Saito, R., Dresselhaus, G.F., Dresselhaus, M.S.: Raman characterization of ABA- and ABC-stacked trilayer graphene. ACS Nano **5**, 8760–8768 (2011)
8. Das, A., Pisana, S., Chakraborty, B., Piscanec, S., Saha, S.K., Waghmare, U.V., Novoselov, K.S., Krishnamurthy, H.R., Geim, A.K., Ferrari, A.C., Sood, A.K.: Monitoring dopants by Raman scattering in an electrochemically top-gated graphene transistor. Nat. Nano **3**, 210–215 (2008)
9. Dean, C.R., Young, A.F., Meric, I., Lee, C., Wang, L., Sorgenfrei, S., Watanabe, K., Taniguchi, T., Kim, P., Shepard, K.L., Hone, J.: Boron nitride substrates for high-quality graphene electronics. Nat. Nano **5**, 722–726 (2010)
10. Xue, J., Sanchez-Yamagishi, J., Bulmash, D., Jacquod, P., Deshpande, A., Watanabe, K., Taniguchi, T., Jarillo-Herrero, P., LeRoy, B.J.: Scanning tunnelling microscopy and spectroscopy of ultra-flat graphene on hexagonal boron nitride. Nat. Mater. **10**, 282–285 (2011)
11. Wang, L., Chen, Z., Dean, C.R., Taniguchi, T., Watanabe, K., Brus, L.E., Hone, J.: Negligible environmental sensitivity of graphene in a hexagonal boron nitride/graphene/h-BN sandwich structure. ACS Nano **6**, 9314–9319 (2012)
12. Li, Z.Q., Henriksen, E.A., Jiang, Z., Hao, Z., Martin, M.C., Kim, P., Stormer, H.L., Basov, D.N.: Dirac charge dynamics in graphene by infrared spectroscopy. Nat. Phys. **4**, 532–535 (2008)
13. Grüneis, A., Attaccalite, C., Pichler, T., Zabolotnyy, V., Shiozawa, H., Molodtsov, S.L., Inosov, D., Koitzsch, A., Knupfer, M., Schiessling, J., Follath, R., Weber, R., Rudolf, P., Wirtz, L., Rubio, A.: Electron–electron correlation in graphite: a combined angle-resolved photoemission and first-principles study. Phys. Rev. Lett. **100**, 037601 (2008)
14. Elias, D.C., Gorbachev, R.V., Mayorov, A.S., Morozov, S.V., Zhukov, A.A., Blake, P., Ponomarenko, L.A., Grigorieva, I.V., Novoselov, K.S., Guinea, F., Geim, A.K.: Dirac cones reshaped by interaction effects in suspended graphene. Nat. Phys. **7**, 701–704 (2011)
15. Luican, A., Li, G., Andrei, E.Y.: Quantized Landau level spectrum and its density dependence in graphene. Phys. Rev. B **83**, 041405 (2011)
16. Attaccalite, C., Rubio, A.: Fermi velocity renormalization in doped graphene. Phys. Status Solidi (b) **246**, 2523–2526 (2009)

17. Trevisanutto, P.E., Giorgetti, C., Reining, L., Ladisa, M., Olevano, V.: Ab initio GW many-body effects in graphene. Phys. Rev. Lett. **101**, 226405 (2008)
18. Chae, J., Jung, S., Young, A.F., Dean, C.R., Wang, L., Gao, Y., Watanabe, K., Taniguchi, T., Hone, J., Shepard, K.L., Kim, P., Zhitenev, N.B., Stroscio, J.A.: Renormalization of the graphene dispersion velocity determined from scanning tunneling spectroscopy. Phys. Rev. Lett. **109**, 116802 (2012)
19. Basko, D.M., Piscanec, S., Ferrari, A.C.: Electron-electron interactions and doping dependence of the two-phonon Raman intensity in graphene. Phys. Rev. B **80**, 165413 (2009)
20. Siegel, D.A., Park, C.-H., Hwang, C., Deslippe, J., Fedorov, A.V., Louie, S.G., Lanzara, A.: Many-body interactions in quasi-freestanding graphene. Proc. Natl. Acad. Sci. **108**, 11365–11369 (2011)
21. Sasaki, K.-I., Kato, K., Tokura, Y., Suzuki, S., Sogawa, T.: Decay and frequency shift of both intervalley and intravalley phonons in graphene: dirac-cone migration. Phys. Rev. B **86**, 201403 (2012)
22. Araujo, P.T., Mafra, D.L., Sato, K., Saito, R., Kong, J., Dresselhaus, M.S.: Phonon self-energy corrections to nonzero wave-vector phonon modes in single-layer graphene. Phys. Rev. Lett. **109**, 046801 (2012)
23. Meric, I., Dean, C., Young, A., Hone, J., Kim, P., Shepard, K.L.: In: Electron Devices Meeting (IEDM), 2010 IEEE International, pp. 23.22.21–23.22.24
24. Pisana, S., Lazzeri, M., Casiraghi, C., Novoselov, K.S., Geim, A.K., Ferrari, A.C., Mauri, F.: Breakdown of the adiabatic Born-Oppenheimer approximation in graphene. Nat. Mater. **6**, 198–201 (2007)
25. Stampfer, C., Molitor, F., Graf, D., Ensslin, K., Jungen, A., Hierold, C., Wirtz, L.: Raman imaging of doping domains in graphene on SiO[sub 2]. Appl. Phys. Lett. **91**, 241907–241903 (2007)
26. Yan, J., Henriksen, E.A., Kim, P., Pinczuk, A.: Observation of anomalous phonon softening in bilayer graphene. Phys. Rev. Lett. **101**, 136804 (2008)
27. Das, A., Chakraborty, B., Piscanec, S., Pisana, S., Sood, A.K., Ferrari, A.C.: Phonon renormalization in doped bilayer graphene. Phys. Rev. B **79**, 155417 (2009)
28. Howard, C.A., Dean, M.P.M., Withers, F.: Phonons in potassium-doped graphene: the effects of electron-phonon interactions, dimensionality, and adatom ordering. Phys. Rev. B **84**, 241404 (2011)
29. Yan, J., Zhang, Y., Kim, P., Pinczuk, A.: Electric field effect tuning of electron–phonon coupling in graphene. Phys. Rev. Lett. **98**, 166802 (2007)
30. Tsang, J.C., Freitag, M., Perebeinos, V., Liu, J., Avouris, P.: Doping and phonon renormalization in carbon nanotubes. Nat. Nano **2**, 725–730 (2007)
31. Piscanec, S., Lazzeri, M., Mauri, F., Ferrari, A.C., Robertson, J.: Kohn anomalies and electron–phonon interactions in graphite. Phys. Rev. Lett. **93**, 185503 (2004)
32. Ando, T.: Anomaly of optical phonon in monolayer graphene. J. Phys. Soc. Jpn. **75**, 124701 (2006)
33. Lazzeri, M., Mauri, F.: Nonadiabatic kohn anomaly in a doped graphene monolayer. Phys. Rev. Lett. **97**, 266407 (2006)
34. Berciaud, S.P., Ryu, S., Brus, L.E., Heinz, T.F.: Probing the intrinsic properties of exfoliated graphene: raman spectroscopy of free-standing monolayers. Nano Lett. **9**, 346–352 (2008)
35. Luo, Z., Cong, C., Zhang, J., Xiong, Q., Yu, T.: Direct observation of inner and outer G[prime] band double-resonance Raman scattering in free standing graphene. Appl. Phys. Lett. **100**, 243107–243104 (2012)
36. Venezuela, P., Lazzeri, M., Mauri, F.: Theory of double-resonant Raman spectra in graphene: intensity and line shape of defect-induced and two-phonon bands. Phys. Rev. B **84**, 035433 (2011)
37. Berciaud, S., Li, X., Htoon, H., Brus, L.E., Doorn, S.K., Heinz, T.F.: Intrinsic line shape of the Raman 2D-mode in freestanding graphene monolayers. Nano Lett. **13**, 3517–3523 (2013)
38. Maultzsch, J., Reich, S., Thomsen, C., Requardt, H., Ordejón, P.: Phonon dispersion in graphite. Phys. Rev. Lett. **92**, 075501 (2004)
39. Lazzeri, M., Attaccalite, C., Wirtz, L., Mauri, F.: Impact of the electron-electron correlation on phonon dispersion: failure of LDA and GGA DFT functionals in graphene and graphite. Phys. Rev. B **78**, 081406 (2008)

40. González, J., Guinea, F., Vozmediano, M.A.H.: Non-fermi liquid behavior of electrons in the half-filled honeycomb lattice (A renormalization group approach). Nucl. Phys. B **424**, 595–618 (1994)
41. González, J., Guinea, F., Vozmediano, M.A.H.: Marginal-Fermi-liquid behavior from two-dimensional Coulomb interaction. Phys. Rev. B **59**, R2474–R2477 (1999)
42. Polini, M., Asgari, R., Barlas, Y., Pereg-Barnea, T., MacDonald, A.H.: Graphene: a pseudochiral fermi liquid. Solid State Commun. **143**, 58–62 (2007)
43. Foster, M.S., Aleiner, I.L.: Graphene via large N: a renormalization group study. Phys. Rev. B **77**, 195413 (2008)
44. Castro Neto, A.H., Guinea, F., Peres, N.M.R., Novoselov, K.S., Geim, A.K.: The electronic properties of graphene. Rev. Mod. Phys. **81**, 109–162 (2009)
45. Hwang, C., Siegel, D.A., Mo, S.-K., Regan, W., Ismach, A., Zhang, Y., Zettl, A., Lanzara, A.: Fermi velocity engineering in graphene by substrate modification. Sci. Rep. **2**, 590 (2012)
46. Mafra, D.L., Samsonidze, G., Malard, L.M., Elias, D.C., Brant, J.C., Plentz, F., Alves, E.S., Pimenta, M.A.: Determination of LA and TO phonon dispersion relations of graphene near the Dirac point by double resonance Raman scattering. Phys. Rev. B **76**, 233407 (2007)
47. Saito, R., Jorio, A., Souza Filho, A.G., Dresselhaus, G., Dresselhaus, M.S., Pimenta, M.A.: Probing phonon dispersion relations of graphite by double resonance Raman scattering. Phys. Rev. Lett. **88**, 027401 (2001)
48. Bostwick, A., Ohta, T., Seyller, T., Horn, K., Rotenberg, E.: Quasiparticle dynamics in graphene. Nat. Phys. **3**, 36–40 (2007)
49. Zhang, Y., Chang, T.-R., Zhou, B., Cui, Y.-T., Yan, H., Liu, Z., Schmitt, F., Lee, J., Moore, R., Chen, Y., Lin, H., Jeng, H.-T., Mo, S.-K., Hussain, Z., Bansil, A., Shen, Z.-X.: Direct observation of the transition from indirect to direct bandgap in atomically thin epitaxial $MoSe_2$. Nat. Nano **9**, 111–115 (2014)

Chapter 3
Coupling of Strongly Localized Graphene Plasmons to Molecular Vibrations

3.1 Introduction

Graphene is a single sheet of carbon atoms arranged in a honeycomb lattice [2–4]. By patterning graphene into nanoribbons, propagating electromagnetic waves in free space can be directly coupled to bound collective charge oscillations, i.e., localized plasmons [5–8]. Strong in-plane light confinement has been revealed experimentally by near-field infrared nanoscopy [7, 8]. However, it still remains elusive how quickly the field decays in the out-of-plane direction. A consequence of the field localization is enhanced light-matter interactions [6, 9–13]. It has already been shown that plasmon resonances in graphene can couple strongly to substrate polar phonons [14, 15] and also to the intrinsic IR-active phonons of bilayer graphene [16]. A closely related, yet unexplored problem is the interaction of plasmons in graphene with the extrinsic vibrational modes in adsorbed species. An experimental demonstration of such a phenomenon, however, is still lacking.

In unpatterned graphene, the electromagnetic boundary conditions and Drude-like conductivity [17] determine the plasmon frequency, which scales with the charge density n and the plasmon wave vector q as $\omega \propto n^{1/4}$ and $\omega \propto q^{1/2}$, respectively [18, 19]. In graphene nanoribbons and other forms of nanostructures, localized plasmons can be directly excited by incident light with zero in-plane momentum, and analogous dispersion relations can be obtained in the quasi-static limit [5], which is appropriate for structures much narrower than the incident light wavelength. These graphene plasmons have resonance frequencies which, depending on their size and doping, lie in the infrared or far-infrared range of the spectrum [15, 20]. Thus, the use of nanoribbons, or other forms of patterned graphene, allows for the efficient and selective coupling of infrared light to surface plasmon modes. In this work, we use infrared transmission spectroscopy to study the interaction of graphene plasmons with vibrational modes in surface-adsorbed thin polymer films. Poly(methyl methacrylate) (PMMA) and poly(vinylpyrrolidone) (PVP) are used as the adsorbates in these experiments. Both polymers contain carbonyl (C=O) groups

Y. Li, *Probing the Response of Two-Dimensional Crystals by Optical Spectroscopy*, Springer Theses, DOI 10.1007/978-3-319-25376-3_3

with vibrational frequencies around 1700 cm^{-1}, but due to their different molecular structures, the exact frequencies are slightly different.

3.2 Sample Preparation and Experiment Setup for Mid-IR Transmission Spectroscopy

The structure of the fabricated samples is shown in Fig. 3.1a. The nanoribbon array is defined by electron beam lithography and etching of graphene grown by chemical vapor deposition (CVD) (Fig. 3.1b). Highly resistive silicon with 280 nm thick surface oxide is used as the substrate that allows light transmission in the mid-infrared range. PMMA layers are spin-coated onto the nanoribbon-covered substrates. The thicknesses of these layers are determined by ellipsometry and controlled using PMMA solutions of different concentrations. Samples with PVP are also prepared by spin-coating, but the thickness of this film is further decreased by washing in copious amounts of water. This dissolves the portion of the polymer that is not in direct contact with the sample surface, leaving a thin PVP residue that is only 2–3 nm thick, as determined by height and phase-contrast AFM. The carbonyl-containing structures of PMMA and PVP are shown in Fig. 3.1c. IR spectra of these

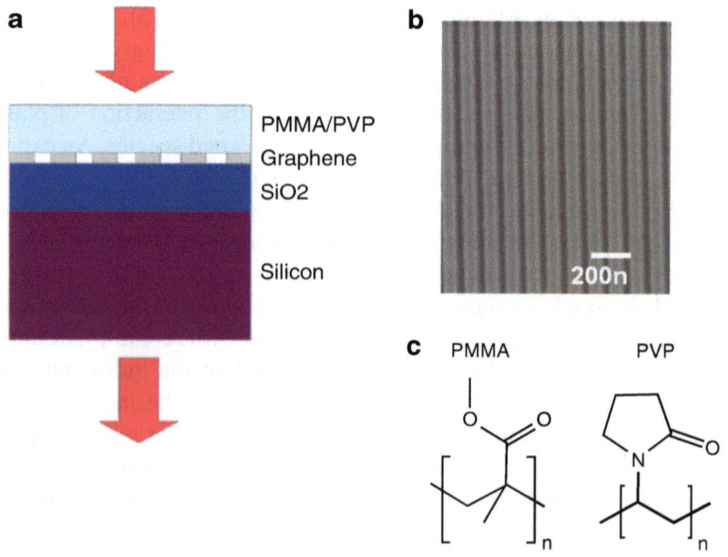

Fig. 3.1 (**a**) Experimental arrangement. Our measurement is conducted in the light transmission configuration. The sample is composed of a thin layer of PMMA or PVP, a graphene nanoribbon (GNR) array, a 280 nm thick SiO_2 layer, and a bulk Si substrate. (**b**) SEM image of graphene nanoribbons. (**c**) PMMA and PVP both contain the carbonyl double-bond (C=O), whose stretching mode has a vibrational frequency around 1700 cm^{-1}

samples are collected in the transmission configuration using an IR microscope coupled to an FTIR spectrometer (Thermo Scientific Nicolet Continuum Infrared Microscope and Nicolet 8700 FTIR spectrometer). The attenuation of the transmission is defined as $A = 1 - T/T_0$ where T is the transmission spectrum with graphene nanoribbons and T_0 is the transmission spectrum of the same system without graphene nanoribbons.

3.3 Determination of the Out-of-Plane Confinement of Graphene Plasmon

The out-of-plane decay of the graphene plasmon field is probed by varying the thickness of the polymer overlayer. Attenuation spectra for a 110 nm wide nanoribbon array with different PMMA overlayer thicknesses are shown in Fig. 3.2a. Here, it is observed that the plasmon resonance strongly red-shifts with increasing PMMA layer thickness, and that the shift saturates for PMMA layer thicknesses larger than about 20 nm. Red shift of the plasmon resonance with an increase in the dielectric constant from the surrounding environment is a well-known behavior of plasmons in metal nanostructures [21]. The saturation of the red-shift provides a direct estimation of the plasmon field decay length into the dielectric. The resonance frequency of the plasmon resonance can be fitted to the exponential function $f = f_0 - \Delta f \exp(-d / d_0)$, where f_0 is the resonance frequency in the limit of a very thick PMMA layer, Δf is the maximum (negative) frequency shift, and d_0 is the

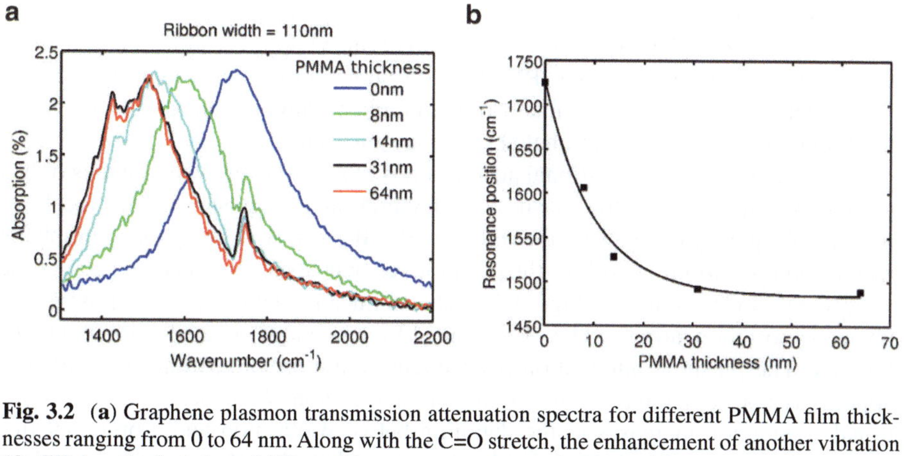

Fig. 3.2 (**a**) Graphene plasmon transmission attenuation spectra for different PMMA film thicknesses ranging from 0 to 64 nm. Along with the C=O stretch, the enhancement of another vibration ($O–CH_3$ bending) at about 1400 cm^{-1} becomes apparent at the larger thicknesses. (**b**) The dependence of the plasmon resonance position on the thickness of the PMMA layer. The exponential fit (*black line*) gives a plasmon field decay length in PMMA of about 10 nm

decay length. From the fitting, we obtain a field decay length of $d_0 = 10\,\text{nm}$ (Fig. 3.2b). Considering that the incident light wavelength is $\lambda_0 \sim 5.8$ µm, this decay length corresponds to a confinement factor in the normal direction of $\lambda_0/d_0 \sim 580$.

In the limit of a thick overlayer, the plasmon field above the graphene surface is completely confined within the PMMA. The observed ~260 cm^{-1} shift of the resonance frequency from bare ribbons to the 64 nm overlayer corresponds to a 15 % decrease in the resonance frequency. An intuitive understanding of this red-shift can be gained by considering the dispersion relation of unpatterned graphene [18]. In this case, the dielectric screening of the plasmon field will result in a reduction of the resonance frequency by 18 % [18], comparable to the measured result for the nanoribbon array. The ~260 cm^{-1} red-shift of the graphene plasmon peak can therefore be understood in terms of dielectric screening from the environment. This shift corresponds to a refractive index sensitivity of ~2650 nm/RIU, a figure of merit used in the plasmon sensor community [22–24].

3.4 Hybridized Plasmon–Phonon Mode and Enhanced IR Sensing

Figure 3.3a shows the attenuation spectra of graphene nanoribbons (90 nm wide) with an 8 nm thick PMMA overlayer for light in resonance with the nanoribbon plasmon and polarized perpendicular and parallel to the ribbons. Here, the polarization dependence of plasmon attenuation is utilized, i.e., plasmons are excited in the nanoribbons when the polarization is perpendicular to the ribbon length, but not when the polarization is parallel [15, 25]. A sharp transparency valley in the attenuation spectra of plasmon resonance is observed at the frequency of the stretching mode of the carbonyl double-bond. The depth of this valley is 0.63 %, while without the graphene the height of the PMMA film attenuation feature is only 0.25 %. Taking into account the filling factor of the graphene ribbons, this corresponds to a fivefold increase in the vibrational absorption sensitivity.

Enhancement of the vibrational absorption is observed even when there is only a partial overlap between the plasmon and vibrational frequencies. The localized plasmon frequency depends on the ribbon width W, $\omega \propto W^{1/2}$ so by using graphene nanoribbons of different widths, the plasmon resonance can be translated across the vibrational mode resonance. Figure 3.3b shows the attenuation spectra of three graphene nanoribbon arrays (110, 120, and 130 nm width) coated with 8 nm of PMMA. The plasmon attenuation peak intensity is normalized and the vibration appears as a peak on the high energy side of the plasmon. We observe that the vibrational attenuation increases as the detuning between plasmon and vibrational frequency decreases ($\Delta I \sim 1/\Delta f$), clearly demonstrating that the enhancement is a resonance effect. This enhancement is a result of the tight in-plane and out-of-plane confinement of electromagnetic field in the graphene plasmon.

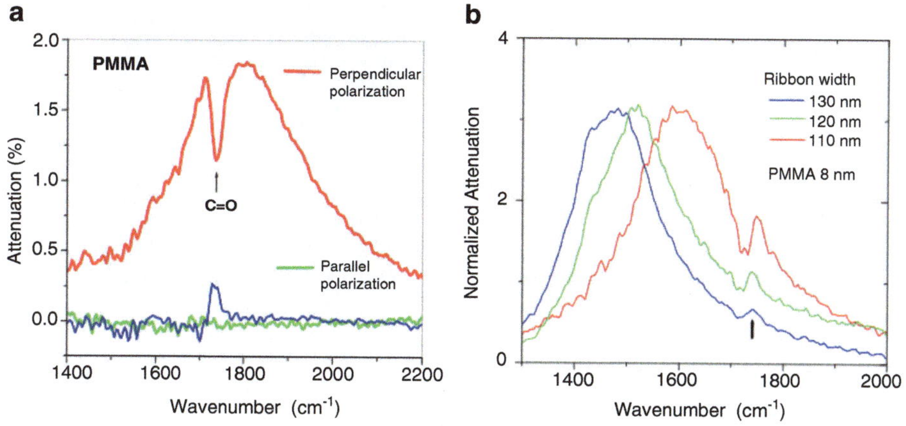

Fig. 3.3 (**a**) Attenuation spectra for light in resonance with the graphene plasmon polarized perpendicular (*red line*) and parallel (*green line*) to the graphene nanoribbons. An 8 nm thick PMMA is spin-coated onto the graphene nanoribbons. Surface plasmons are excited in the perpendicular polarization case, but not excited in the parallel polarization case. The attenuation spectrum of an 8 nm thick PMMA layer (without nanoribbons) is shown as the *blue line*. (**b**) Attenuation spectra of three graphene nanoribbon arrays (110, 120, and 130 nm width) coated with 8 nm of PMMA. The plasmon attenuation peak intensity is normalized and the vibration appears as a peak at the high energy side of the plasmon. The two excitations overlap only partially

3.5 Finite Difference Time Domain Simulation and a Phenomenological Model

To understand the origin of the IR absorption enhancement, we performed finite-difference time-domain (FDTD) simulations in which the device structure is set to be the same as in Fig. 3.1a. The plasmon is excited in the nanoribbon by a plane wave whose frequency is in resonance with the plasmon, and periodic boundary conditions are used to simulate the array of nanoribbons. The conductivity of graphene is assumed to follow a Drude form. Furthermore, the DC conductivity and the scattering rate in the Drude form determine the resonance frequency and the width of the plasmon peaks, respectively. In Fig. 3.4 we show results for two different plasmon widths (damping rates) $\gamma = 50\ \text{cm}^{-1}$ and $\gamma = 250\ \text{cm}^{-1}$. The enhancement in the field intensity is expressed by taking the ratio of the electric field intensities with and without the nanoribbons at each spatial point, $|E|^2/|E_0|^2$. As can be seen from the color-coded Fig. 3.4, large enhancements in the field intensity can be generated (more than three orders of magnitude) near the edges of the nanoribbons. From the same plot, we obtain an out-of-plane field decay length of ~10 nm, in agreement with our experimental measurement discussed in Sect. 3.3.

This enhancement in the local electric field is the source of the observed enhancement in the interaction between light and the carbonyl vibration in the polymer film.

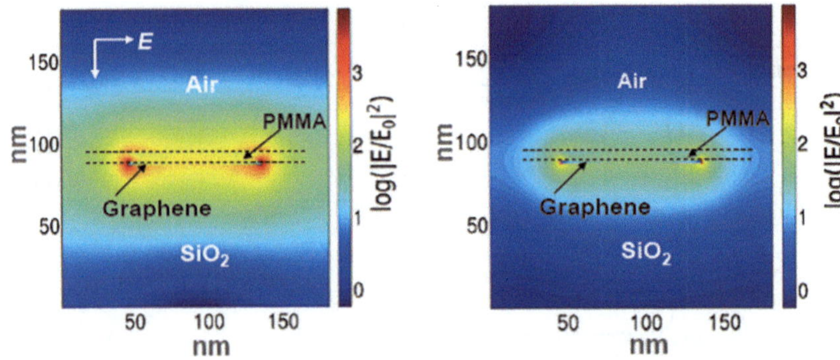

Fig. 3.4 Spatial enhancement of electric field intensity near a graphene nanoribbon. Finite-difference time-domain (FDTD) simulations of the field enhancement in the vicinity of nanoribbons for two values of plasmon damping, $\gamma = 50\ \text{cm}^{-1}$ (*left*) and $\gamma = 250\ \text{cm}^{-1}$ (*right*)

The enhancement can be potentially further increased through improved graphene sample quality (reduced plasmon damping), and by the use of stacked graphene layers [20]. These optimization procedures deserve further work, but are beyond the scope of the present study.

Figure 3.5 illustrates the variation of the vibrational spectrum as a function of the detuning between the plasmon and the vibrational excitation. To understand qualitatively the variation in intensity and line shape, we model the system using a pair of coupled electrical dipole oscillators [26]. In this model, the plasmon in graphene and carbonyl vibration in the polymer are represented by two dipole oscillators that are coupled to each other. Within the quasi-static approximation, the relations between the physical parameters in this model are summarized in the following coupled differential equations:

$$\ddot{x}_1 + \gamma_1 \dot{x}_1 + \omega_1^{\ 2} x_1 = A_1 e^{i\omega t} - C_{12} x_2 \tag{3.1}$$

$$\ddot{x}_2 + \gamma_2 \dot{x}_2 + \omega_2^{\ 2} x_2 = -C_{21} x_1 \tag{3.2}$$

Here, x_1 and x_2 denote the electric polarizations corresponding to the graphene plasmon and the carbonyl vibrational mode response to the plasmon, γ_1 and γ_2 denote the decay rates of the graphene plasmon and the vibrational mode, ω_1 and ω_2 are their respective resonant frequencies, C_{12} and C_{21} are the coupling constants between the two dipoles, and A_1 is proportional to the amplitude of the driving electric field. The total transmission attenuation of a thin sheet, which is proportional to the real part of the conductivity in the weak response limit [27] can then be obtained. The vibrational frequency of the carbonyl group ω_2 is fixed and the other parameters are varied. As shown in Fig. 3.5 (gray line), this phenomenological model successfully describes the observed evolution of the spectrum, from nearly symmetric

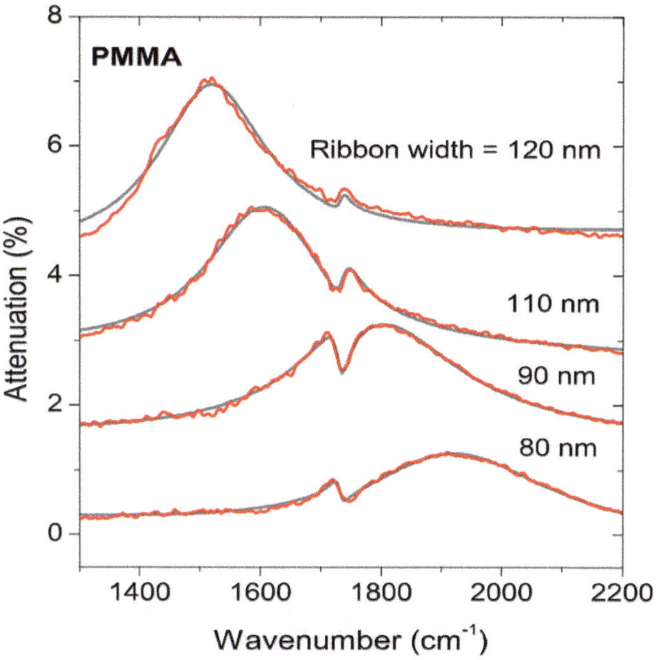

Fig. 3.5 Graphene plasmon extinction spectra for different graphene nanoribbon widths. The changes in intensity, line shape, and the development of a transmission transparency near the 1700 cm^{-1} C=O vibration are due to the interaction of the graphene plasmon and the molecular vibrations in PMMA. The *gray lines* are phenomenological fits using the coupled electrical dipole oscillator model (see text)

resonances to asymmetric peaks (Fano line shape) and the transmission transparency. We note that the evolution of the spectrum is also produced by direct FDTD simulations, confirming the coupling mechanism to be of electromagnetic nature.

To demonstrate the specificity of the enhanced light-matter interaction, PVP was also studied. Height and phase-contrast AFM analysis of the PVP-coated nanoribbon array reveals that the maximum thickness of the PVP film is 3 nm. Figure 3.6 shows the spectrum of graphene nanoribbons covered by this overlayer. Similar to the PMMA case, a phonon induced transparency valley appears within the plasmon resonance peak. Detection of this feature in such a thin and nonuniform film further demonstrates the plasmon-induced enhancement capability of the nanoribbon array. Moreover, due to the 50 cm^{-1} difference in the carbonyl double-bond vibrational peak frequencies of PMMA and PVP, the locations of the corresponding transparency valleys also differ by the same amount. This demonstrates the molecular specificity of this sensing technology, even without surface functionalization.

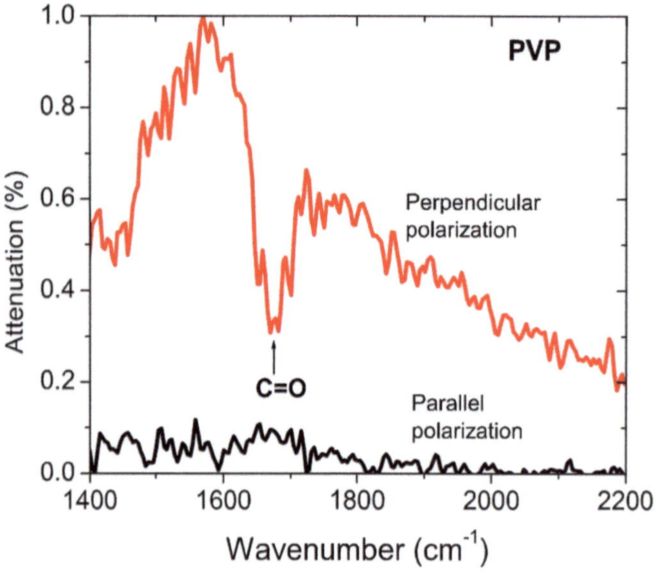

Fig. 3.6 Attenuation spectra for graphene nanoribbons covered with a PVP layer with a thickness <3 nm, for perpendicular (*red line*) and parallel (*black line*) polarizations

3.6 Conclusion

In conclusion, we have measured the out-of-plane decay length of graphene plasmon to be 10 nm from the peak shift due to dielectric screening of adsorbed layers. Considering a free space wavelength of 5.8 μm, the decay length corresponds to an out-of-plane confinement factor of ~580. The strong out-of-plane confinement, in addition to the strong in-plane confinement, results in significant enhancement in the near-field electromagnetic field intensity and hence enhanced light-matter interaction. This is demonstrated through an induced transparency in the graphene plasmon spectral peak by coupling to the vibrational mode of the surface-adsorbed polymers. The depth of the induced transparency is five times larger in magnitude comparing to that of the absorbance peak by the vibrational mode alone. As the graphene plasmon frequency is tuned away from the vibrational mode frequency, the induced transparency gradually evolves into a Fano-like line shape and eventually turns back into an absorption peak when it fully separates from the plasmon resonance in graphene. Through FDTD simulations and a phenomenological coupled oscillator model, we show that the coupling of the vibrational modes in adsorbed molecules and graphene plasmon is of electromagnetic nature. Lastly, as a sensing technique, the plasmon-vibrational mode coupling exhibits molecular specificity, i.e., different molecular compounds can be differentiated from the peak frequencies of the vibrational mode induced transparencies.

References

1. Li, Y., Yan, H., Farmer, D.B., Meng, X., Zhu, W., Osgood, R.M., Heinz, T.F., Avouris, P.: Graphene plasmon enhanced vibrational sensing of surface-adsorbed layers. Nano Lett. **14**, 1573–1577 (2014)
2. Novoselov, K.S., Geim, A.K., Morozov, S.V., Jiang, D., Katsnelson, M.I., Grigorieva, I.V., Dubonos, S.V., Firsov, A.A.: Two-dimensional gas of massless Dirac fermions in graphene. Nature **438**, 197–200 (2005)
3. Zhang, Y., Tan, Y.-W., Stormer, H.L., Kim, P.: Experimental observation of the quantum Hall effect and Berry's phase in graphene. Nature **438**, 201–204 (2005)
4. Geim, A.K., Novoselov, K.S.: The rise of graphene. Nat. Mater. **6**, 183–191 (2007)
5. Ju, L., Geng, B., Horng, J., Girit, C., Martin, M., Hao, Z., Bechtel, H.A., Liang, X., Zettl, A., Shen, Y.R., Wang, F.: Graphene plasmonics for tunable terahertz metamaterials. Nat. Nano **6**, 630–634 (2011)
6. Koppens, F.H.L., Chang, D.E., García de Abajo, F.J.: Graphene plasmonics: a platform for strong light-matter interactions. Nano Lett **11**, 3370–3377 (2011)
7. Chen, J., Badioli, M., Alonso-Gonzalez, P., Thongrattanasiri, S., Huth, F., Osmond, J., Spasenovic, M., Centeno, A., Pesquera, A., Godignon, P., Zurutuza Elorza, A., Camara, N., de Abajo, F.J.G., Hillenbrand, R., Koppens, F.H.L.: Optical nano-imaging of gate-tunable graphene plasmons. Nature **487**, 77–81 (2012)
8. Fei, Z., Rodin, A.S., Andreev, G.O., Bao, W., McLeod, A.S., Wagner, M., Zhang, L.M., Zhao, Z., Thiemens, M., Dominguez, G., Fogler, M.M., Neto, A.H.C., Lau, C.N., Keilmann, F., Basov, D.N.: Gate-tuning of graphene plasmons revealed by infrared nano-imaging. Nature **487**, 82–85 (2012)
9. Liu, G.-B., Shan, W.-Y., Yao, Y., Yao, W., Xiao, D.: Three-band tight-binding model for monolayers of group-VIB transition metal dichalcogenides. Phys. Rev. B **88**, 085433 (2013)
10. Luk'yanchuk, B., Zheludev, N.I., Maier, S.A., Halas, N.J., Nordlander, P., Giessen, H., Chong, C.T.: The Fano resonance in plasmonic nanostructures and metamaterials. Nat. Mater. **9**, 707–715 (2010)
11. Adato, R., Yanik, A.A., Amsden, J.J., Kaplan, D.L., Omenetto, F.G., Hong, M.K., Erramilli, S., Altug, H.: Ultra-sensitive vibrational spectroscopy of protein monolayers with plasmonic nanoantenna arrays. Proc. Natl. Acad. Sci. **106**, 19227–19232 (2009)
12. Neubrech, F., Pucci, A., Cornelius, T.W., Karim, S., García-Etxarri, A., Aizpurua, J.: Resonant plasmonic and vibrational coupling in a tailored nanoantenna for infrared detection. Phys. Rev. Lett. **101**, 157403 (2008)
13. Halas, N.J., Lal, S., Chang, W.-S., Link, S., Nordlander, P.: Plasmons in strongly coupled metallic nanostructures. Chem. Rev. **111**, 3913–3961 (2011)
14. Fei, Z., Andreev, G.O., Bao, W., Zhang, L.M., McLeod, A.S., Wang, C., Stewart, M.K., Zhao, Z., Dominguez, G., Thiemens, M., Fogler, M.M., Tauber, M.J., Castro-Neto, A.H., Lau, C.N., Keilmann, F., Basov, D.N.: Infrared nanoscopy of Dirac plasmons at the graphene–SiO_2 interface. Nano Lett. **11**, 4701–4705 (2011)
15. Yan, H., Low, T., Zhu, W., Wu, Y., Freitag, M., Li, X., Guinea, F., Avouris, P., Xia, F.: Damping pathways of mid-infrared plasmons in graphene nanostructures. Nat. Photon. **7**, 394–399 (2013)
16. Yan, H., Low, T., Guinea, F., Xia, F., Avouris, P.: Tunable phonon-induced transparency in bilayer graphene nanoribbons. Nano Lett. **14**, 4581–4586 (2014)
17. Horng, J., Chen, C.-F., Geng, B., Girit, C., Zhang, Y., Hao, Z., Bechtel, H.A., Martin, M., Zettl, A., Crommie, M.F., Shen, Y.R., Wang, F.: Drude conductivity of Dirac fermions in graphene. Phys. Rev. B **83**, 165113 (2011)
18. Jablan, M., Buljan, H., Soljačić, M.: Plasmonics in graphene at infrared frequencies. Phys. Rev. B **80**, 245435 (2009)
19. Hwang, E.H., Das Sarma, S.: Dielectric function, screening, and plasmons in two-dimensional graphene. Phys. Rev. B **75**, 205418 (2007)

20. Yan, H., Li, X., Chandra, B., Tulevski, G., Wu, Y., Freitag, M., Zhu, W., Avouris, P., Xia, F.: Tunable infrared plasmonic devices using graphene/insulator stacks. Nat. Nano **7**, 330–334 (2012)
21. Wang, F., Shen, Y.R.: General properties of local plasmons in metal nanostructures. Phys. Rev. Lett. **97**, 206806 (2006)
22. Homola, J., Yee, S.S., Gauglitz, G.: Surface plasmon resonance sensors: review. Sens. Actuators B **54**, 3–15 (1999)
23. Liedberg, B., Nylander, C., Lunström, I.: Surface plasmon resonance for gas detection and biosensing. Sens. Actuators **4**, 299–304 (1983)
24. Zhao, Y., Hu, X., Chen, G., Zhang, X., Tan, Z., Chen, J., Ruoff, R.S., Zhu, Y., Lu, Y.: Infrared biosensors based on graphene plasmonics: modeling. Phys. Chem. Chem. Phys. **15**, 17118–17125 (2013)
25. Freitag, M., Low, T., Zhu, W., Yan, H., Xia, F., Avouris, P.: Photocurrent in graphene harnessed by tunable intrinsic plasmons. Nat. Commun. **4**, 1951 (2013)
26. Garrido Alzar, C.L., Martinez, M.A.G., Nussenzveig, P.: Classical analog of electromagnetically induced transparency. Am. J. Phys. **70**, 37–41 (2002)
27. Tinkham, M.: Energy gap interpretation of experiments on infrared transmission through superconducting films. Phys. Rev. **104**, 845–846 (1956)

Chapter 4
Dielectric Response of a Thin Sheet

4.1 Introduction to the Dielectric Response of a Material

The response of a charged free particle to an electromagnetic field is given by the combined effect of the Coulomb force $\vec{F}_E = q\vec{E}$ and the Lorentz force $\vec{F}_B = q\vec{v} \times \vec{B}$, where q denotes the charge of the particle, $\vec{v}$ denotes its velocity, and $\vec{E}$ and $\vec{B}$ denote the electric and magnetic fields. If the velocity of the electron is much smaller than the speed of light c, the Coulomb force dominates the Lorentz force by a factor of $c / v_{\perp}$ where c is the speed of light and $v_{\perp}$ is the electron velocity component perpendicular to the magnetic field. For an electron in a crystal system, velocity is not a strictly well-defined quantum number; however, the qualitative picture still applies for a Bloch wave packet. We thus conclude that the electromagnetic response of a material can be described as a pure electrical response with sufficient accuracy in most laboratory scenarios.

The *macroscopic* electrical response, or the dielectric response, of a material originates from the averaged field-induced modifications to the *microscopic* quantum states [1]. In the dipole approximation, the dielectric response to an applied electric field $\vec{E}$ is solely captured by the macroscopic electrical polarization $\vec{P}$ induced in the material.

4.2 Linear Dielectric Response

In the situation where the applied electric field is much smaller than the internal field of a molecule, it is often convenient and of sufficient accuracy to consider the electrical polarization that is linear in the applied electric field—the linear dielectric response. Four of the most commonly used physical quantities to describe the linear dielectric response are the polarizability χ, the dielectric function ε, the refractive index n, and the conductivity σ. Each of them provides complete information of the

Y. Li, *Probing the Response of Two-Dimensional Crystals by Optical Spectroscopy*, Springer Theses, DOI 10.1007/978-3-319-25376-3_4

Table 4.1 Material properties used to describe the dielectric response

Material properties	Definitions
$\chi = \chi_1 + i\chi_2$	$\chi = P/E$
$\varepsilon = \varepsilon_1 + i\varepsilon_2$	$\varepsilon = 1 + \chi$
$n = n_1 + in_2$	$n = \sqrt{1+\chi}$
$\sigma = \sigma_1 + i\sigma_2$	$\sigma = -i\varepsilon_0\omega\chi$

dielectric response and is in general a complex quantity to account for the phase difference between the applied field and the induced polarization. We provide the definitions of χ in terms of E and P, and the definition of ε, n, and σ in terms of χ in Table 4.1. Implicitly, these quantities are in general tensors relating the polarization and electric field vectors and are functions of the frequency.

4.3 Nonlinear Dielectric Response

As the strength of the applied electric field increases, the dielectric response deviates from the simple linear response. The polarization can be written as a Taylor expansion of the incident electric field:

$$P_i = \chi_{ij}^{(1)} E_j + \chi_{ijk}^{(2)} E_j E_k + \cdots. \tag{4.1}$$

The first term gives the linear dielectric response. The second term corresponds to the second-order nonlinear response, which gives rise to the second-harmonic generation process discussed later in this thesis.

Historically, the study of nonlinear optical properties of materials is enabled by the invention of laser, which produces much brighter electromagnetic radiation than conventional light sources. The advent of pulsed lasers, in particular the mode-locked femtosecond lasers, further increased the peak electric field, which boosts the nonlinear optical signal. With a focused Ti:sapphire laser beam, one can readily achieve an electric field strength of ~10^8 V/m (at a wavelength of about 800 nm for Ti:sapphire femtosecond mode-locked laser). Considering the internal field of a solid state system, which is on the order of 10^9 V/m, measurable nonlinear response is anticipated. Since nonlinear process produces response at frequency different from the excitation frequency in general, the nonlinear response, even smaller comparing to the excitation, can be clearly distinguished from the linear response by spectrally dispersing the signal with a grating.

4.4 Models for the Dielectric Response of a Thin Film

There are two most commonly used models to describe the optical response of a thin film. The first model treats the thin layer as a uniform material with a volumetric dielectric function. For an atomically thin layer, an effective thickness of the layer needs to be defined, which is taken naturally as the interlayer spacing in the bulk material.

The second model, the 2D sheet model, does not require the definition of the effective thickness. In this approximation, the dielectric response of the thin film is described by a complex sheet conductivity σ^s, which relates the induced sheet current to the applied electrical field. The criterion for the validity of the second approximation is that the optical thickness of the film is much smaller than the wavelength, i.e., $\sqrt{|\varepsilon|}d \ll \lambda$ [2]. Important optical responses of a thin film can be written in terms of the sheet conductivity in rather concise formulas in terms of σ^s. As an example, the absorbance is given by:

$$A = \frac{4Z_0\sigma_1^s}{\left|1+n_s+Z_0\sigma^s\right|^2}, \tag{4.2}$$

where Z_0 is the impedance vacuum, $\sigma_1{}^s$ is the real part of the complex sheet conductivity, and n_s is the refractive index of the substrate.

A further approximation is used when the dielectric response of the material is weak, by which we mean that $Z_0\sigma^s \ll 1$ (or equivalently $|\varepsilon|d \ll \lambda$). In this limit, the absorbance formula can be expressed to the linear order in its Taylor expansion with good accuracy:

$$A = \frac{4Z_0\sigma_1^s}{\left|1+n_s\right|^2}. \tag{4.3}$$

In the case of graphene and monolayer TMDCs, in the optical range, λ/d is on the order of 10^3 and $|\varepsilon|$ is in the range of 0–50. This renders the 2D sheet model almost always a satisfactory description for single to few-layer TMDCs. However, the linearized version can produce significant error when the magnitude of the dielectric function is large.

In Fig. 4.1, we present the absorbance spectra of monolayer MoS_2 obtained from the three models using the dielectric function inferred from our optical measurement (Chap. 5). Quantitatively, the thin slab and the 2D sheet models give almost identi-

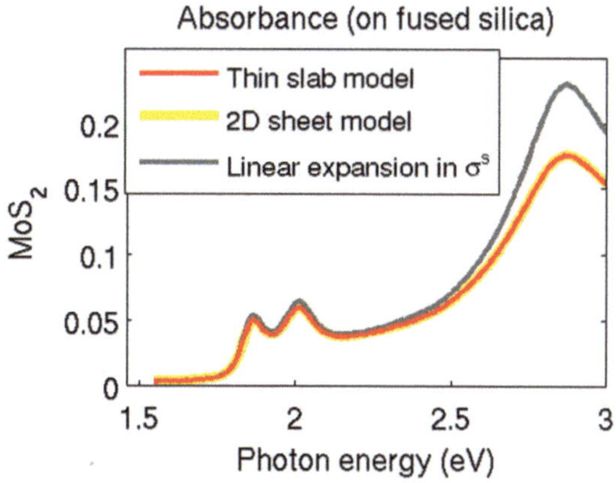

Fig. 4.1 Absorbance spectra obtained using the thin slab model (*red*), the 2D sheet model (*yellow*), and the linearized 2D sheet model (*black*), using the same dielectric functions for MoS_2 measured in our experiment (Chap. 5)

cal absorbance spectra, since the condition $\sqrt{|\varepsilon|} \ll \lambda / 2\pi d$ is satisfied very well. The linear expansion in $Z_0\sigma^s$ produces reasonable absorbance value throughout the whole spectral range, while the deviation from the other two approximation schemes becomes more noticeable in the blue region where the material response is strong.

References

1. Jackson, J.D.: Classical Electrodynamics. Wiley, New York (1998)
2. Tinkham, M.: Energy gap interpretation of experiments on infrared transmission through superconducting films. Phys. Rev. **104**, 845–846 (1956)

Chapter 5
Measurement of the Optical Dielectric Function of Monolayer Transition Metal Dichalcogenides: MoS_2, $MoSe_2$, WS_2, and WSe_2

5.1 Introduction

Transition metal dichalcogenide (TMDC) crystals have emerged as a new class of semiconductors that display distinctive properties at monolayer thickness [2–4]. Their optical properties are of particular interest and importance. They exhibit a transition to direct bandgap semiconductors at monolayer thickness [5, 6], offer access to valley degree of freedom by optical helicity [3, 7, 8], and display strong excitonic properties, with tightly bound neutral and charged excitons, as well as a non-hydrogenic Rydberg series of excited states [9–11]. The materials thus provide an excellent testing ground for the physics of 2D systems and many-body effects in solids. At the same time, they have attracted much interest for applications in optoelectronics as light emitters, detectors, and photovoltaic devices [12–14].

The most basic description of light-matter interactions in TMDC monolayers is given by the materials' complex dielectric function. The dielectric function provides a meeting point between experiment and theories of excited-state properties of this novel class of materials; knowledge of the dielectric function is also crucial for the characterization of these materials and their use in emerging applications. Despite its central role, the optical dielectric functions for these materials have yet to be reported.

In this chapter, we present a systematic measurement and analysis of the complex in-plane dielectric functions of four monolayer TMDCs—MoS_2, $MoSe_2$, WS_2, and WSe_2—over the photon energies between 1.5 and 3 eV. We obtain the dielectric functions by Kramers–Kronig constrained analysis of the reflectance spectra. These data also yields the absorption spectra for the monolayers. The strong light-matter interaction leads to peaks in the imaginary part of the dielectric function for the A exciton exceeding 30 in some of the materials, with a corresponding single-layer absorption exceeding 15 %. For the purpose of further understanding of the reso-

Y. Li, *Probing the Response of Two-Dimensional Crystals by Optical Spectroscopy*, Springer Theses, DOI 10.1007/978-3-319-25376-3_5

nance features in the dielectric response, we extracted the oscillator strengths, as well as transition energies of the resonance peaks and analyzed several trends, which are identified to arise from excitonic effects.

5.2 Sample Preparation for Reflectance Measurement

We prepared monolayer TMDC samples on fused silica substrates by mechanical exfoliation [15] and chemical vapor deposition [16–20]. The monolayer samples were first identified by their optical contrast under a microscope and then verified by Raman and photoluminescence spectroscopy. The dielectric functions were determined by reflectance measurements of the samples on the transparent substrate at room temperature. For this purpose, we made use of the broadband emission from a tungsten halogen lamp, which was spatially filtered by a pinhole before being focused onto the sample using an objective. The spot size on sample was about 2 μm. The reflected light was collected by the same objective and deflected by a beam splitter to a spectrometer equipped with a CCD camera cooled to liquid-nitrogen temperature. We determined the reflectance spectra for the samples on fused silica substrates by normalizing the measured reflected power by that from the bare fused silica substrate. The reflectance from the fused silica substrate was calibrated using literature values for the material's refractive index [21]. The spectral resolution of our measurements was 2 meV, much narrower than any feature in the spectra. In our experimental configuration, the optical fields lie in the plane of the sample, thus only the *in-plane* dielectric response is accessed.

5.3 Measurement of Reflectance of the Monolayer TMDCs

Absolute reflectance spectra for the TMDC monolayers on fused silica are presented in Fig. 5.1a–d. For all four TMDC monolayers, the two lowest energy peaks in the reflectance spectra correspond to the excitonic features associated with interband transitions at the K (K′) point in the Brillouin zone [22]. The two features, denoted by A and B, are attributed to the splitting of the valence band by spin-orbit coupling [23]. At higher energies we observe the spectrally broad response from the higher-lying interband transitions [22], including the transitions near the Γ point [24, 25].

The data presented in this chapter correspond to representative TMDC monolayer flakes. Different samples of the same material generally exhibit very close dielectric response (Fig. 5.2a). However, charge doping, strain, and inhomogeneity

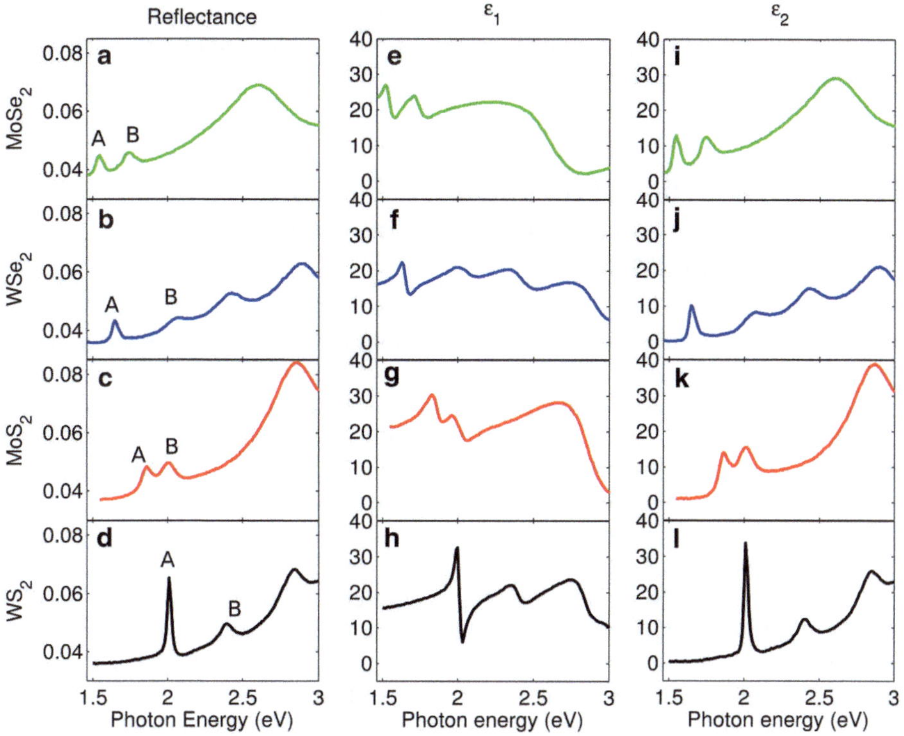

Fig. 5.1 Optical response of monolayers of $MoSe_2$, WSe_2, MoS_2, and WS_2 exfoliated on fused silica: (**a–d**) The measured reflectance spectra. (**e–h**) The real part of the dielectric function, ε_1. (**i–l**) The imaginary part of the dielectric function, ε_2. The peaks labeled A and B in (**a–d**) correspond to excitons from the two spin-orbit split transitions at the K-point of the Brillouin zone

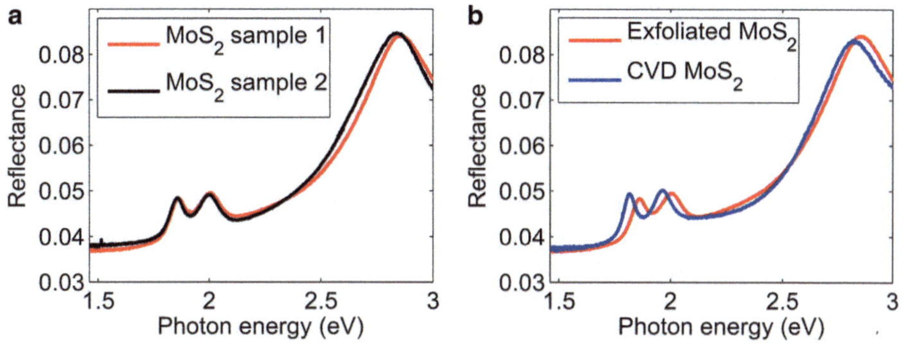

Fig. 5.2 (**a**) Comparison of the reflectance contrast spectra of two different exfoliated MoS_2 monolayers. (**b**) Comparison of the reflectance spectra of exfoliated (*red*) and CVD-grown (*blue*) MoS_2 monolayers

may cause slight changes in the position and line widths of the narrow excitonic peaks in the spectral response, as illustrated by the comparison of exfoliated and CVD-grown samples presented in Fig. 5.2b.

5.4 Determination of the Dielectric Function from Reflectance Spectrum Using Kramers–Kronig Constrained Analysis

We obtain the dielectric function $\varepsilon(E)$ for the four different TMDC monolayers by fitting a modeled complex dielectric function to the experimental reflectance spectra. The optimized dielectric functions yield reflectance spectra that match experimental spectra within the thickness of lines in Fig. 5.1a–d. In Fig. 5.1e–l, we show the resultant real and imaginary parts of the dielectric functions for $MoSe_2$, WSe_2, MoS_2, and WS_2 over the spectral range of 1.5 eV $\leq E \leq$ 3 eV.

For the purposes of the optical model, we treat the monolayer as a homogeneous medium with an effective thickness given by the interlayer spacing of the respective bulk material [22]. The optical reflectance is calculated by the standard thin-film analysis [26], which fully accounts for all interference effects. We model the complex dielectric function $\varepsilon(E) = \varepsilon_1(E) + i\varepsilon_2(E)$ the samples as a function of photon energy E using a superposition of Lorentzian oscillators:

$$\varepsilon(E) = 1 + \sum_{k=1}^{N} \frac{f_k}{E_k{}^2 - E^2 - iE\gamma_k}. \tag{5.1}$$

Here f_k and γ_k are the oscillator strength and the linewidth of the kth oscillator, and E_k runs over the full spectral range. In our modeling, we choose a uniform spacing between the oscillators of 2 meV, much smaller than the narrowest feature in the optical spectra, and a full width of the individual oscillators of 10 meV to produce a spectrally smooth response. Satisfaction of the Kramers–Kronig relation is built into the functional form of (5.1) and does not need to be considered separately. We fit the measured reflectance data using the dielectric function of (5.1). The oscillator strengths f_k between 1.5 and 3 eV are varied, while the parameters for photon energies $E > 3$ eV are fit to the bulk values.

Our reflectance measurements only cover the spectral range of 1.5 eV $\leq E \leq$ 3 eV. In general, optical transitions outside this energy range need to be considered for an accurate determination of ε_1, since the non-resonant tails of higher or lower lying spectral features may contribute to the range of interest. In the infrared, the dielectric response arises from polar phonons and free carriers. Based on the weakness of the contribution of polar phonons in bulk materials [27, 28] to the visible dielectric response, we omit this term. The influence of free carriers on the dielectric functions in the optical frequency range is estimated to be less than 0.1, using the Drude model for a carrier density of 10^{12} cm^{-2}, chosen on the basis of transport characteristics of usual TMDC transistor devices [5, 29]. We therefore also neglect this contribution. The influence of higher-energy electronic transitions, however, needs to be taken into account; strong transitions lie just beyond our measurement window. For this purpose, we use data for the bulk materials [30, 31], including transitions up to $E = 30$ eV. As discussed below, the transition energies shift on the order of several hundred meV compared with transition energies above 3 eV. We therefore expect the bulk dielectric function to provide an adequate approximation of the dielectric function to correct for the off-resonant response of higher-lying transitions.

Rather than applying the Kramers–Kronig constrained analysis, one could consider measurement of both reflection and transmission spectra of the sample. In principle, this yields two independent measurements for each photon energy, thus directly determining ε_1 and ε_2, without recourse to the Kramers–Kronig analysis. For a thin layer on a transparent substrate, as is the case here, however, both the transmission and reflection spectra are dominated by ε_2. Consequently, this approach cannot be applied reliably.

5.5 Surface Conductivity and Absorbance

We can also express the material response in terms of the optical conductivity $\sigma(E) = \sigma_1(E) + i\sigma_2(E) = -i(\varepsilon_0 E / \hbar)[\varepsilon(E) - 1]$. We present this result in the form of the complex sheet conductivity $\sigma^s(E)$, where d denotes the layer thickness. For a layer in which there is no significant propagation effect for light passing through the material, i.e., $|\varepsilon(E) - 1|^{\frac{1}{2}}(E / \hbar c)d \ll 1$ the sheet conductivity provides a full description of the optical response. In Fig. 5.3a–d, we plot the real part of the sheet conductivity, $\sigma_1^s(E)$, for the four TMDCs.

In Fig. 5.3e–h we present the absorption spectra for free-standing monolayers of the four TMDC materials based on the measured dielectric functions. The overall absorbance for all four materials in this frequency range is on the order of 10 %, demonstrating strong light-matter interaction even for a monolayer.

5.6 Extraction of the A and B Peaks Splitting

To extract quantitative properties of the excitonic resonances, we parameterize ε_2 using a small, but physically meaningful number of Lorentzians terms, allowing for an extraction of both the exciton peak energies and the relative oscillator strengths. The energy differences between the A and B excitonic transitions are shown in Fig. 5.4 for the monolayers [30, 31]. Here, the four materials, as well as data from recently exfoliated $MoTe_2$ monolayer [32], are presented as a function

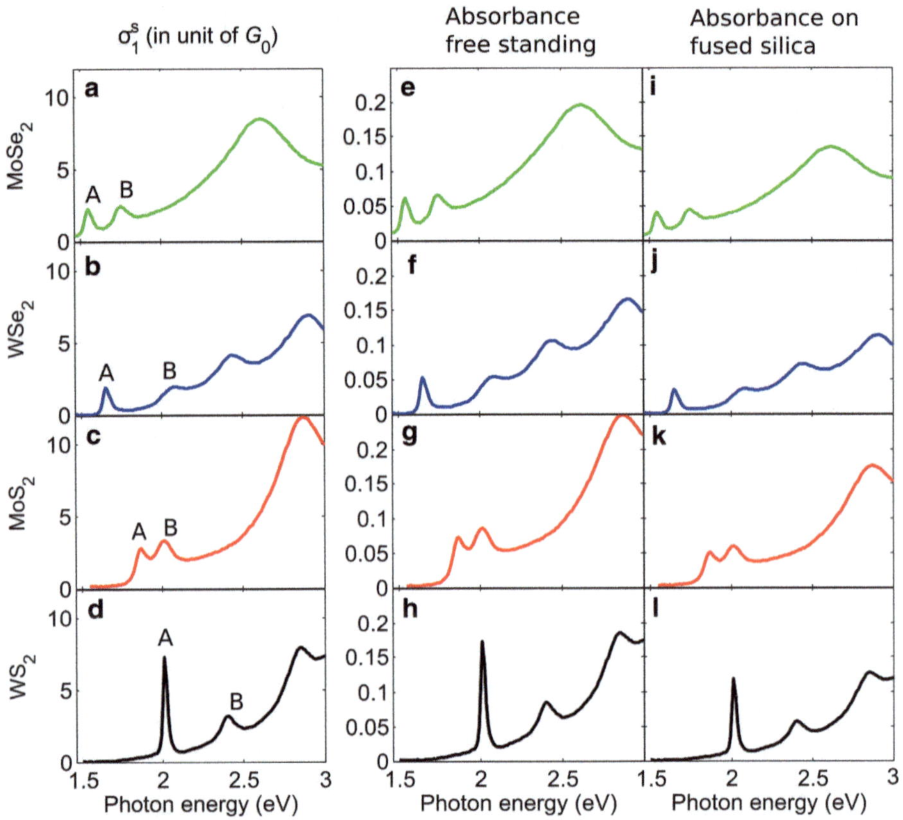

Fig. 5.3 Real part of the sheet conductivity (in units of $G_0 = 2e^2/h$) and absorbance for (free-standing and supported) monolayers of (**a, e, i**) MoSe2, (**b, f, j**) WSe2, (**c, g, k**) MoS_2, and (**d, h, l**) WS_2. The peaks labeled A and B correspond to excitons from the two spin-orbit split transitions at the K-point of the Brillouin zone

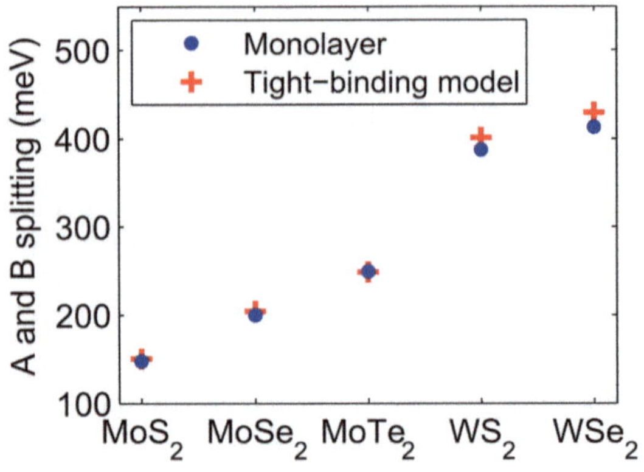

Fig. 5.4 Energy difference between the A and B transitions (*blue dots*) in monolayer $MoSe_2$, WSe_2, $MoTe_2$, MoS_2, and WS_2. The result for $MoTe_2$ is taken from [32]. The predicted splittings (*red crosses*) within a three-band tight-binding model [37] are shown for comparison

of the effective atomic number of the material, determined by weighting each element's relative contributions to the spin-orbit coupling [33]. The monolayer results correspond well with the previously reported data for MoS_2, WS_2, and WSe_2 monolayers [5, 34]; the predicted A–B splitting from a three-band tight-binding model [35] is also shown in Fig. 5.4. The good agreement between our experiment and the predictions of the tight-binding model (fit to density functional calculations) is surprising at first sight, since the band calculation neglects Coulomb interactions. In general, however, the exciton binding energy correction to the transition energy is largely offset by the band self-energy when the Coulomb interactions are taken into account.

5.7 Comparison of the Monolayer and Bulk Dielectric Function

The dielectric functions for the monolayer TMDC crystals obtained in this work can be compared with the dielectric functions for the corresponding bulk materials [30, 31] (Fig. 5.5). While the dielectric functions for monolayer and bulk TMDCs show an overall similarity, differences in the spectral responses are readily seen. Qualitatively, there is a general broadening of the resonance features in the bulk materials comparing to the monolayers, due to the extra relaxation channels from interlayer coupling.

The resonance energies in the monolayer dielectric function are generally shifted from the corresponding bulk material. In Fig. 5.6, we present the energy difference between the spectral peaks in the imaginary part of the monolayer and bulk dielectric functions. We plot the A peak energy differences for MoS_2, $MoSe_2$, WS_2, and WSe_2, the C peaks energy differences for MoS_2, $MoSe_2$, and WS_2, and D peak energy difference for WSe_2. We chose the D peak instead of the C peak for WSe_2 since the C peak in bulk WSe_2 overlaps the B peak, hindering an accurate determination of the peak energy. Comparing the bulk to the monolayer, we see a much stronger red shift for the C/D peak than for the A peak. As discussed in the previous section, the *net* Coulomb contribution to the transition energies is expected to be reduced significantly because of the cancellation of electron–electron and electron–hole interactions. In considering the physical origin of the shift of the peaks from the monolayer to the bulk, we consequently focus on quantum-confinement effects. The A peak corresponds to the excitons located at the K-point in the Brillouin zone, which are composed primarily of the metal *d*-orbitals. Thus the A exciton is spatially localized in the metal atom plane and interlayer interactions in the bulk material do not lower the transition energy significantly. The C/D feature is, however, associated with transitions away from the K-point and involves a significant contribution from the chalcogen orbitals [24, 33, 36]. We therefore expect appreciably stronger interlayer interactions in building up the bulk material from

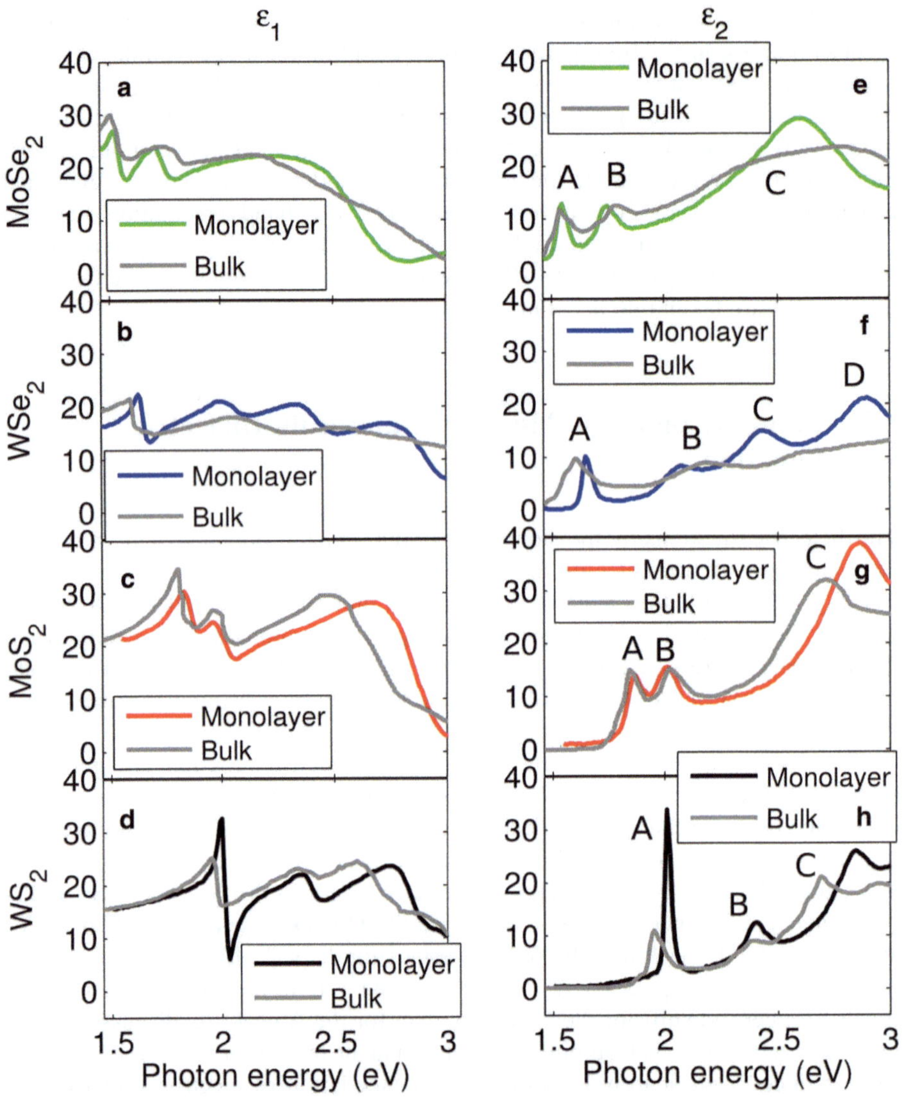

Fig. 5.5 Comparison of the monolayer TMDC dielectric function (colored, reproduced from Fig. 1) with that of the corresponding bulk material (*gray*) for (**a**, **e**) MoS_2, (**b**, **f**) $MoSe_2$, (**c**, **g**) WS_2, and (**d**, **h**) WSe_2. The peaks in ε_2 are labeled A, B, C, and D

monolayers. This provides a physical rationale for the larger observed red shift of the C/D transition compared to that of the A exciton in comparing the bulk material to the monolayer. The behavior of the C/D transition with increasing layer thickness is analogous to the quantum confinement effect of the lowest indirect optical transition, which involves states at Γ-point [37].

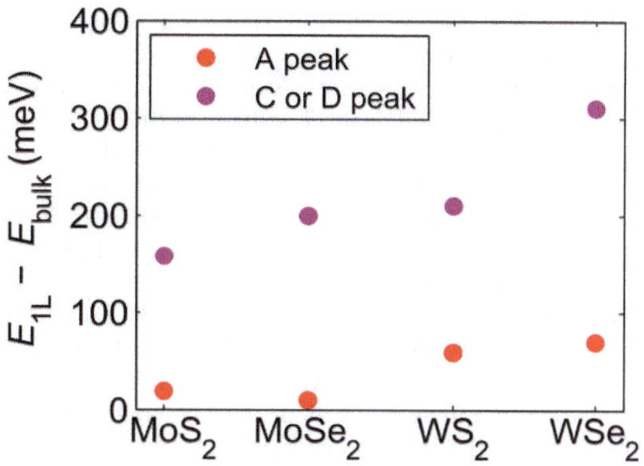

Fig. 5.6 Energy shift of the peaks in the dielectric function to the peak positions in respective bulk materials. The A peak corresponds to the band-edge exciton at the K-point in the Brillouin zone and the C or D peak (depending on the material system, as discussed in the text) corresponds to the transitions away from the K-point

5.8 Conclusion

In conclusion, we have experimentally determined the complex dielectric function of monolayer MoS_2, $MoSe_2$, WS_2, and WSe_2 over photon energies between 1.5 and 3.0 eV. The dielectric functions imply strong light-matter interactions in even monolayers of TMDCs, with a peak absorbance of each of the four materials exceeding 15 %. Comparing to the bulk materials, we observed only a slight red shift of the A feature, but a more significant shift of the higher-lying (C/D) features. This behavior can be understood as a reflection of the different orbital character and interlayer interactions of the wavefunctions relevant for the lower and higher energy transitions.

References

1. Li, Y., Chernikov, A., Zhang, X., Rigosi, A., Hill, H.M., van der Zande, A.M., Chenet, D.A., Shih, E.-M., Hone, J., Heinz, T.F.: Measurement of the optical dielectric function of monolayer transition-metal dichalcogenides: MoS_2, $MoSe_2$, WS_2, and WSe_2. Phys. Rev. B **90**, 205422 (2014)
2. Butler, S.Z., Hollen, S.M., Cao, L., Cui, Y., Gupta, J.A., Gutiérrez, H.R., Heinz, T.F., Hong, S.S., Huang, J., Ismach, A.F., Johnston-Halperin, E., Kuno, M., Plashnitsa, V.V., Robinson, R.D., Ruoff, R.S., Salahuddin, S., Shan, J., Shi, L., Spencer, M.G., Terrones, M., Windl, W., Goldberger, J.E.: Progress, challenges, and opportunities in two-dimensional materials beyond graphene. ACS Nano **7**, 2898–2926 (2013)
3. Cao, T., Wang, G., Han, W., Ye, H., Zhu, C., Shi, J., Niu, Q., Tan, P., Wang, E., Liu, B., Feng, J.: Valley-selective circular dichroism of monolayer molybdenum disulphide. Nat. Commun. **3**, 887 (2012)
4. Xu, X., Yao, W., Xiao, D., Heinz, T.F.: Spin and pseudospins in layered transition metal dichalcogenides. Nat. Phys. **10**, 343–350 (2014)
5. Mak, K.F., Lee, C., Hone, J., Shan, J., Heinz, T.F.: Atomically thin MoS_2: a new direct-gap semiconductor. Phys. Rev. Lett. **105**, 136805 (2010)
6. Splendiani, A., Sun, L., Zhang, Y., Li, T., Kim, J., Chim, C.-Y., Galli, G., Wang, F.: Emerging photoluminescence in monolayer MoS_2. Nano Lett. **10**, 1271–1275 (2010)

7. Mak, K.F., He, K., Shan, J., Heinz, T.F.: Control of valley polarization in monolayer MoS_2 by optical helicity. Nat. Nano **7**, 494–498 (2012)
8. Zeng, H., Dai, J., Yao, W., Xiao, D., Cui, X.: Valley polarization in MoS_2 monolayers by optical pumping. Nat. Nano **7**, 490–493 (2012)
9. Chernikov, A., Berkelbach, T.C., Hill, H.M., Rigosi, A., Li, Y., Aslan, O.B., Reichman, D.R., Hybertsen, M.S., Heinz, T.F.: Exciton binding energy and nonhydrogenic rydberg series in monolayer WS_2. Phys. Rev. Lett. **113**, 076802 (2014)
10. He, K., Kumar, N., Zhao, L., Wang, Z., Mak, K.F., Zhao, H., Shan, J.: Tightly bound excitons in monolayer WSe_2. Phys. Rev. Lett. **113**, 026803 (2014)
11. Mak, K.F., He, K., Lee, C., Lee, G.H., Hone, J., Heinz, T.F., Shan, J.: Tightly bound trions in monolayer MoS_2. Nat. Mater. **12**, 207–211 (2013)
12. Pospischil, A., Furchi, M.M., Mueller, T.: Solar-energy conversion and light emission in an atomic monolayer *p–n* diode. Nat. Nano **9**, 257–261 (2014)
13. Baugher, B.W.H., Churchill, H.O.H., Yang, Y., Jarillo-Herrero, P.: Optoelectronic devices based on electrically tunable *p–n* diodes in a monolayer dichalcogenide. Nat. Nano **9**, 262–267 (2014)
14. Ross, J.S., Klement, P., Jones, A.M., Ghimire, N.J., Yan, J., Mandrus, D.G., Taniguchi, T., Watanabe, K., Kitamura, K., Yao, W., Cobden, D.H., Xu, X.: Electrically tunable excitonic light-emitting diodes based on monolayer WSe_2 *p–n* junctions. Nat. Nano **9**, 268–272 (2014)
15. Novoselov, K.S., Jiang, D., Schedin, F., Booth, T.J., Khotkevich, V.V., Morozov, S.V., Geim, A.K.: Two-dimensional atomic crystals. Proc. Natl. Acad. Sci. U.S.A. **102**, 10451–10453 (2005)
16. Lee, Y.-H., Zhang, X.-Q., Zhang, W., Chang, M.-T., Lin, C.-T., Chang, K.-D., Yu, Y.-C., Wang, J.T.-W., Chang, C.-S., Li, L.-J., Lin, T.-W.: Synthesis of large-area MoS_2 atomic layers with chemical vapor deposition. Adv. Mater. **24**, 2320–2325 (2012)
17. Zhan, Y., Liu, Z., Najmaei, S., Ajayan, P.M., Lou, J.: Large-area vapor-phase growth and characterization of MoS_2 atomic layers on a SiO_2 substrate. Small **8**, 966–971 (2012)
18. Liu, K.-K., Zhang, W., Lee, Y.-H., Lin, Y.-C., Chang, M.-T., Su, C.-Y., Chang, C.-S., Li, H., Shi, Y., Zhang, H., Lai, C.-S., Li, L.-J.: Growth of large-area and highly crystalline MoS_2 thin layers on insulating substrates. Nano Lett. **12**, 1538–1544 (2012)
19. Lu, X., Lin, J., Utama, M.I., Gong, X., Zhang, J., Zhao, Y., Pantelides, S.T., Wang, J., Dong, Z., Liu, Z., Zhou, W., Xiong, Q.: Large-area synthesis of monolayer and few-layer MoSe2 films on SiO2 substrates. Nano Lett. **14**, 2419–2425 (2014)
20. Cong, C., Shang, J., Wu, X., Cao, B., Peimyoo, N., Qiu, C., Sun, L., Yu, T.: Synthesis and optical properties of large-area single-crystalline 2D semiconductor WS2 monolayer from chemical vapor deposition. Adv. Opt. Mater. **2**, 131–136 (2014)
21. Bass, M., DeCusatis, C., Enoch, J., Lakshminarayanan, V., Li, G., MacDonald, C., Mahajan, V., Van Stryland, E.: Handbook of Optics, Third Edition Volume IV: Optical Properties of Materials, Nonlinear Optics, Quantum Optics (set). McGraw-Hill Education, New York (2009)
22. Wilson, J.A., Yoffe, A.D.: The transition metal dichalcogenides discussion and interpretation of the observed optical, electrical and structural properties. Adv. Phys. **18**, 193–335 (1969)
23. Mattheiss, L.F.: Band structures of transition-metal-dichalcogenide layer compounds. Phys. Rev. B **8**, 3719–3740 (1973)
24. Qiu, D.Y., da Jornada, F.H., Louie, S.G.: Optical spectrum of MoS2: many-body effects and diversity of exciton states. Phys. Rev. Lett. **111**, 216805 (2013)
25. Carvalho, A., Ribeiro, R.M., Castro Neto, A.H.: Band nesting and the optical response of two-dimensional semiconducting transition metal dichalcogenides. Phys. Rev. B **88**, 115205 (2013)
26. Hecht, E.: Optics. Addison-Wesley Longman, New York (2002)
27. Lucovsky, G., White, R.M., Benda, J.A., Revelli, J.F.: Infrared-reflectance spectra of layered group-IV and group-VI transition-metal dichalcogenides. Phys. Rev. B **7**, 3859–3870 (1973)
28. Wieting, T.J., Verble, J.L.: Infrared and Raman studies of long-wavelength optical phonons in hexagonal MoS2. Phys. Rev. B **3**, 4286–4292 (1971)
29. Radisavljevic, B., Radenovic, A., Brivio, J., Giacometti, V., Kis, A.: Single-layer MoS2 transistors. Nat. Nano **6**, 147–150 (2011)
30. Beal, A.R., Hughes, H.P.: Kramers-Kronig analysis of the reflectivity spectra of 2H-MoS 2, 2H-MoSe 2 and 2H-MoTe 2. J. Phys. C **12**, 881 (1979)

31. Beal, A.R., Liang, W.Y., Hughes, H.P.: Kramers-Kronig analysis of the reflectivity spectra of 3R-WS 2 and 2H-WSe 2. J. Phys. C **9**, 2449 (1976)
32. Ruppert, C., Aslan, O.B., Heinz, T.F.: Optical properties and band gap of single- and few-Layer MoTe2 crystals. Nano Lett. **14**, 6231–6236 (2014)
33. Zhu, Z.Y., Cheng, Y.C., Schwingenschlögl, U.: Giant spin-orbit-induced spin splitting in two-dimensional transition-metal dichalcogenide semiconductors. Phys. Rev. B **84**, 153402 (2011)
34. Zhao, W., Ghorannevis, Z., Chu, L., Toh, M., Kloc, C., Tan, P.-H., Eda, G.: Evolution of electronic structure in atomically thin sheets of WS2 and WSe2. ACS Nano **7**, 791–797 (2012)
35. Liu, G.-B., Shan, W.-Y., Yao, Y., Yao, W., Xiao, D.: Three-band tight-binding model for monolayers of group-VIB transition metal dichalcogenides. Phys. Rev. B **88**, 085433 (2013)
36. Coehoorn, R., Haas, C., de Groot, R.A.: Electronic structure of MoSe2, MoS2, and WSe2. II. The nature of the optical band gaps. Phys. Rev. B **35**, 6203–6206 (1987)
37. Zhao, W., Ribeiro, R.M., Toh, M., Carvalho, A., Kloc, C., Castro Neto, A.H., Eda, G.: Origin of indirect optical transitions in few-layer MoS2, WS2, and WSe2. Nano Lett. **13**, 5627–5634 (2013)

References

[illegible]

Chapter 6
Measurement of the Second-Order Nonlinear Susceptibility and Probing Symmetry Properties of Few-Layer MoS_2 and h-BN by Optical Second-Harmonic Generation

6.1 Introduction

In atomically layered materials, the individual layers will generally exhibit different symmetry from the corresponding bulk crystals. Few-layer materials of different layer thickness can moreover have distinct symmetries from one another, even when their thickness differs by only one atomic layer. Since symmetry plays a critical role in defining the material properties, such differences in symmetry lead to significant differences in material properties. Among the layered materials, few-layer graphene, MoS_2, and hexagonal BN have attracted great interest due to their distinctive properties and potential for novel applications. Monolayer graphene is a semi-metal without a bandgap. However, in bilayer graphene, a sizable bandgap can be opened when inversion symmetry is lifted by an out-of-plane electric field [2–4]. Single-layer MoS_2 and h-BN are non-centrosymmetric materials, while their bilayers and bulk counterparts are expected to exhibit inversion symmetry. The broken inversion symmetry in transition metal dichalcogenides of single-layer thickness has been demonstrated to permit the production of long-lived valley polarization by optical helicity [5–7]. It has also been predicted and very recently shown that the non-centrosymmetric single-layer sheets exhibit a strong intrinsic piezoelectric response [8–13].

Bulk MoS_2 is built layers consisting of two hexagonal lattices of S atoms and a sheet of Mo atoms occupying trigonal prismatic sites between the S sheets (Fig. 6.1a). Bulk h-BN is formed from layers of honeycomb lattices of B and N atoms, occupying, respectively, the A and B sublattices in each layer (Fig. 6.1b). Both MoS_2 and h-BN layers are stacked in 2H order, with pairs of layers forming a unit that is repeated along the *c*-axis [14, 15]. Because of their similar structure, bulk MoS_2 and h-BN crystals belong to the same centrosymmetric D_{6h}^4 space group. However, few-layer segments of the bulk crystal will generally exhibit different symmetry properties. In the case of few-layer MoS_2 and h-BN, slices of even layer thickness belong to the centrosymmetric D_{3d}^3 space group, while slices of odd layer

Y. Li, *Probing the Response of Two-Dimensional Crystals by Optical Spectroscopy*, Springer Theses, DOI 10.1007/978-3-319-25376-3_6

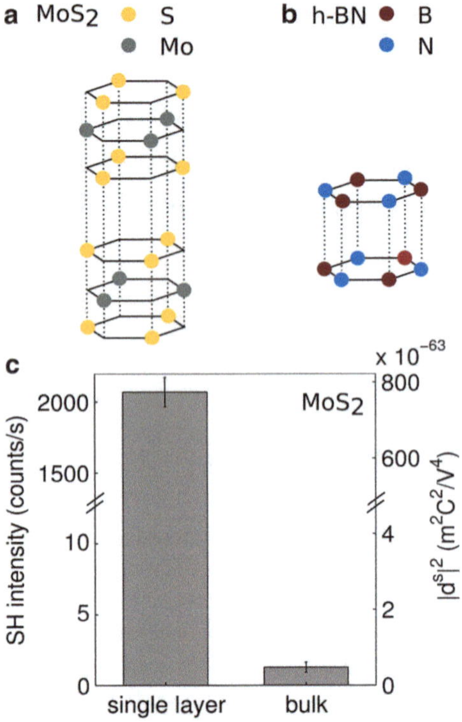

Fig. 6.1 (**a, b**): Structure of repeating two-layer units in 2H-stacked MoS_2 and h-BN. (**c**) The SH intensity and surface nonlinear susceptibility for single-layer MoS_2 and from the surface of bulk MoS_2. The vertical axes are broken to allow the display of the very different SH response from the single layer and bulk material

thickness belong to the non-centrosymmetric D^1_{3h} space group. It has been demonstrated that stable few-layer units of both of these materials can be isolated by mechanical exfoliation [16]. Since the surface of a bulk crystal can undergo lattice reconstruction, whether these isolated atomically thin sheets exhibit the varying crystallographic symmetries expected from their bulk parents is a question of fundamental scientific interest.

In this work, we probe the variations in symmetry in few-layer MoS_2 and h-BN by optical second-harmonic generation (SHG) [17, 18]. The SHG process, which is described in the electric-dipole approximation by a third-rank tensor, is known to be a sensitive probe of symmetry of surface layers [19–21]. Using this method, we verify that inversion symmetry is indeed broken in MoS_2 and h-BN samples of odd layer thickness. This is manifested by the large enhancement of the second-harmonic (SH) response of these samples compared both to samples of even numbers of layers and to the bulk crystals [22]. The threefold rotational symmetry of the layered materials (for odd numbers of layers) was probed by the orientational dependence of SH intensity. In this manner, we establish SHG as a precise optical probe of the orientation of the crystallographic axes. A recent study clearly demonstrated the orientational dependence of SHG from single-layer MoS_2 and applied this sensitivity to imaging crystal domains in monolayer MoS_2 samples grown by chemical vapor deposition [23]. Our findings further indicate that the isolated few-layer MoS_2 and h-BN retain the crystallographic structure of the layers in the bulk crystals. For odd-layer MoS_2 and h-BN samples, we further determine the absolute strength of the SH susceptibility and study its evolution with layer thickness. To describe the different behavior observed for the two material systems with layer thickness, we introduce a model accounting for the phase shift and absorption of the optical fields. In MoS_2, the interlayer electronic coupling and the corresponding change in electronic structure as a function of layer thickness are found to be of importance in defining the linear and nonlinear response.

6.2 Experimental Setup for Optical SHG Measurement

We prepared few-layer MoS_2 and h-BN samples on fused silica substrates by mechanical exfoliation [16]. The layer thicknesses of the different exfoliated MoS_2 samples were determined by the central wavelength and intensities of the photoluminescence from the direct gap (for monolayers) and from indirect bandgap transition (for thicker samples) [24, 25]. For analysis of the h-BN samples, we relied on the relative intensity of the Raman peak of E_{2g} optical phonon mode near 1370 cm^{-1} [26], confirming our assignments of layer thickness by measurements with atomic-force microscopy (AFM).

The SH measurements were performed in a reflection geometry using normal incidence excitation. The pump radiation was supplied by a mode-locked Ti:sapphire oscillator operating at an 80-MHz repetition rate. The pulses were of 90-fs duration and centered at a wavelength of 810 nm. Using a ×100 objective, we focused the pump radiation to a spot size of about 1 μm on the sample. For measurements of MoS_2, we limited the average laser power to 1 mW; for measurements of h-BN, laser powers up to 10 mW power were employed. The retroreflected SH signal was collected by the same objective, separated by a beam splitter, and filtered to block reflected fundamental radiation. An analyzer was used to select the polarization components of the SH radiation lying either parallel or perpendicular to the polarization of the pump beam. The SH signal was detected by a thermoelectrically cooled CCD camera after it was dispersed in a spectrometer. The SH character of the detected radiation was verified by its wavelength and quadratic power dependence on the pump intensity. In our setup, we could freely rotate the samples to obtain the orientational dependence of the SH response.

6.3 Probing the Absence of Inversion Symmetry and the Crystal Rotational Symmetry by Polarization Resolved SHG

The dramatic role of symmetry can be seen in the contrast of the SH response of a monolayer of MoS_2 and of the bulk MoS_2 crystal. We observed an increase in the reflected SH intensity by more than a thousand times for the monolayer compared

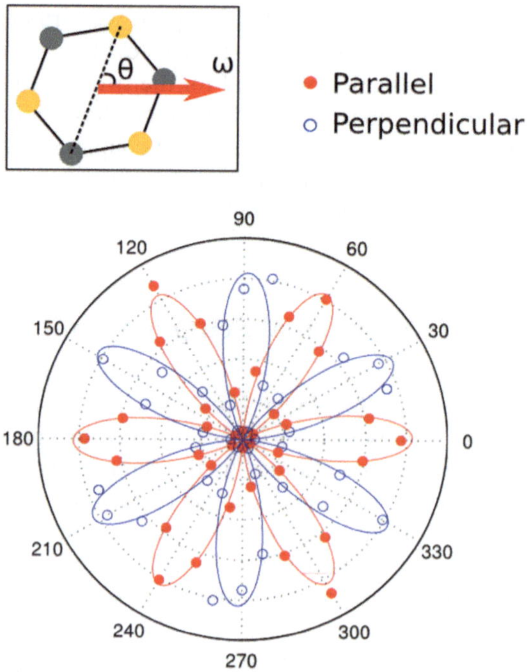

Fig. 6.2 Polar plot of the SH intensity from single-layer MoS_2 as a function of the crystal's azimuthal angle θ. The SH radiation components detected parallel (*red*) and perpendicular (*blue*) to the polarization of the fundamental field are shown. The symbols are experimental data and the *solid lines* are fits to the symmetry analysis described in the text. The angle $\theta = 0$ corresponds to an orientation where the polarization of the pump beam lies along a mirror plane of the crystal, as shown in the *inset*

to the surface of the bulk material (Fig. 6.1c). This enhancement reflects the fact that bulk MoS_2 is centrosymmetric, and previous studies have shown it to have a very weak SH response [22]. The origin for the small, but finite SH signal from bulk MoS_2 is discussed below.

For both the parallel and perpendicular polarization components, the SH intensity from single-layer MoS_2 exhibited a strongly varying, sixfold symmetric response as a function of azimuthal angle for rotation about its surface normal (Fig. 6.2) [8]. Similar results were obtained for single-layer h-BN samples. As discussed above, on the basis of the structure of the bulk crystal, exfoliated monolayers of both MoS_2 and h-BN are expected to belong to the D_{3h}^{1} symmetry group. For this case, we expect only one independent non-vanishing element of the nonlinear response: $d = d_{yyy} = -d_{yxx} = -d_{xxy} = -d_{xyx}$ [27]. This symmetry information determines the dependence of the two polarization components of SH response on sample orientation:

$$I_{\parallel} = I_0 \cos^2(3\theta) \tag{6.1}$$

$$I_{\perp} = I_0 \sin^2(3\theta) \tag{6.2}$$

Here θ denotes the angle between the mirror plane in the crystal structure and the polarization of the pump beam, and I_0 is the maximum intensity of the SH response. As shown in Fig. 6.2, the experimental data agree well with these predicted expressions for the angular variation. The same orientational dependence of the SH intensity was also observed in MoS_2 and h-BN with 1, 3, 5 layers, precisely as expected based on the identical space group for all of these materials [28]. We note that the dependence of SH electric field (not the SH intensity) on crystallographic orientation actually exhibits threefold rotational symmetry, in accordance with the lattice symmetry.

6.4 Measurement of the Second-Order Nonlinear Susceptibility

We describe the second-order nonlinear response from the various single- and few-layer samples in terms of a surface nonlinear susceptibility tensor d^s that relates the applied laser field to the induced nonlinear sheet polarization in the sample. Calibrated values of the second-order nonlinear coefficients for few-layer MoS_2 and h-BN samples were obtained by comparing the corresponding SH intensity I_s to the reference SH intensity I_r from the surface of a z-cut bulk crystal of α-quartz. For this measurement, the quartz crystal was oriented so that its largest SH coefficient (d_{11}) was accessed. We then determined the surface second-order nonlinear response d^s from the ratio I_s/I_r using the relation

$$\frac{d^s}{d_{11}} = \frac{c}{4\omega\left[n(\omega)+n(2\omega)\right]}\sqrt{\frac{I_s}{I_r}}. \tag{6.3}$$

Here ω is the fundamental (angular) frequency, and $n(\omega)$ and $n(2\omega)$ represent the (ordinary) refractive indices of crystalline quartz at the fundamental and SH frequencies, respectively. Equation (6.3) is derived by taking the ratio of an analytic expression for the SH intensity produced by a 2D polarizable sheet on a semi-infinite dielectric substrate [29] to that produced by the boundary of a semi-infinite nonlinear medium with a bulk second-order response [30]. It should be noted that the surface nonlinear response is expected to be a complex quantity for the (resonant) MoS_2 samples. We did not determine the phase of the response, and d^s should be understood as the modulus of the complex surface nonlinear susceptibility.

From the intensity maxima in the previous measurement and known SH nonlinear susceptibility for quartz of $d_{11}=2.7\times10^{-24}$ C/V^2 (7.4×10^{-10} esu) [31], we obtain for single-layer MoS_2 and h-BN values of the surface nonlinear susceptibility of $d^s_{MoS2}=8.8\times10^{-31}$ m·C/V^2 (2.4×10^{-14} esu) and $d^s_{h\text{-}BN}=3.0\times10^{-32}$m·C/V^2 ($8.3\times10^{-16}$ esu). The effective volume second-order nonlinear susceptibilities for MoS_2 and h-BN are then $d_{MoS2}=1.4\times10^{-21}$ C/V^2 and $d_{h\text{-}BN}=9.1\times10^{-23}$ C/V^2. Here we have assumed effective thicknesses of the layers of $t_{MoS2}=0.62$ nm [14] and $t_{h\text{-}BN}=0.33$ nm [15], corresponding to the layer separation in the bulk materials. The effective nonlinear susceptibility of h-BN is seen to be comparable to that of common transparent nonlinear crystals such as $LiNbO_3$ and β-BaB_2O_4. Because of resonant enhancement, the MoS_2 monolayer has an effective nonlinear susceptibility that is one order of magnitude larger, similar to that of absorbing nonlinear crystals such as GaAs and Te.

6.5 Probing Inversion Symmetry Breaking by SHG Intensity

For few-layer h-BN, strong SH radiation with similar intensity occurs for odd numbers of layers, while even numbers of layers do not show measurable SHG (Fig. 6.3a). The intensity values for different layer thicknesses were taken at the lattice orientations corresponding to the peak SH response. The signal for even numbers of layers is at most 2 % of that emitted from the single layer. This large suppression of SH radiation in even numbers of layers shows that these structures do indeed exhibit inversion symmetry, as expected from the bulk crystal.

To understand the thickness dependence of the SHG response more quantitatively, we introduce a simple model of the expected response based on electronically decoupled layers, but that still accounts for electromagnetic propagation effects at both the fundamental and SH frequencies. In computing phase shifts and absorption, we assume that the thin slabs to have the same linear dielectric constants as their bulk counterparts [32, 33]. In addition to propagation through the sample material, we also include the influence of the reflection of both the fundamental and SH fields from the underlying substrate. The SH radiation from each layer is calculated taking into account the alternating sign of the SH susceptibility in adjacent layers (Fig. 6.3b). As shown in Fig. 6.3a, the model captures the dramatic reduction SHG in samples of even layer thickness. For two layers, for example, the model predicts a SH response that is just 0.008 % of that from a monolayer. The nonzero result arises from the small, but finite optical phase shift between the layers, which acts to prevent complete cancellation. This effect corresponds formally to inclusion of nonlinear response beyond the electric-dipole approximation in orders of spatial

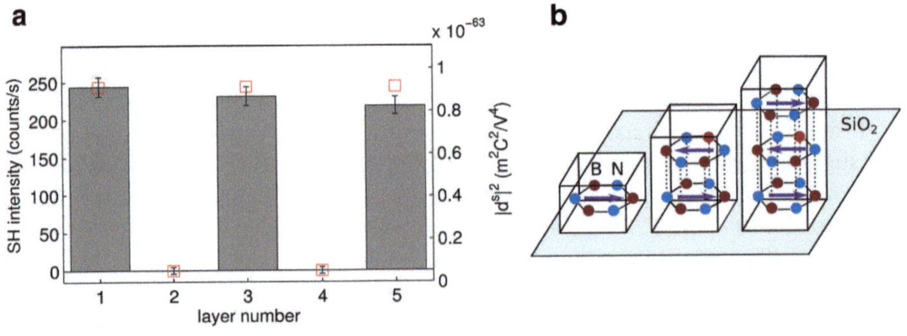

Fig. 6.3 **(a)** The layer dependence of SH intensity and surface nonlinear susceptibility for few-layer h-BN. The histogram shows the experimental results, with uncertainties indicated by the error bars. The *red squares* indicate the results of a model (described in the text) that considers the phase evolution of the fundamental and SH fields, but neglects changes in the electronic structure from interlayer coupling. The model is calibrated to the results for the single-layer sample. (b) Illustration of the model for the calculation of the SH intensity as a function of layer thickness. The individual layers of h-BN are approximated as thin slabs and the polarization is assumed to be generated at the center of each layer. The alternating orientation of successive layers leads to an alternation in the nonlinear susceptibility of each layer

non-locality. For odd layer numbers N, in the simplest picture, the signal from a block of $N-1$ layers fully cancels out, leaving the residual signal from just one layer. Thus, no thickness dependence would be expected. The aforementioned propagation effects, in the transparent h-BN system, modify this expectation only very slightly. The predicted signal from five layers is, for example, 99.8 % of that from a monolayer, in reasonable agreement with the experimental results.

For the case of MoS_2, we again observed strong enhancement of the SH intensity in odd numbers of layers compared to the response for even numbers of layers. The SH intensity from odd numbers of layers, however, decreases significantly from one layer to five layers (Fig. 6.4a), in contrast to the nearly constant response for h-BN. Since the bandgap of MoS_2 is smaller than the SH photon energy, we expect light absorption to play a significant role. We applied the model described above to determine the expected influence of these propagation effects, including attenuation (Fig. 6.4a). The results are in agreement with the experimental trends for the variation of the SH intensity with layer thickness. The model, however, underestimates the decrease in SHG actually observed with increasing layer thickness. The additional reduction can be attributed to interlayer coupling. These effects modify the electronic structure of few-layer MoS_2 and induce further changes in the linear and nonlinear susceptibilities. The evolving electronic structure with layer number for few-layer MoS_2 is demonstrated by reflectivity contrast measurements (Fig. 6.4b). In particular, the change of resonance feature close to 2ω can have a substantial effect on the both linear response and SH susceptibility of few-layer MoS_2 [27]. In the limit of a bulk crystal, the strong, but imperfect cancellation of SHG from neighboring layers leads to a weak, but finite SH response. This response is expected even

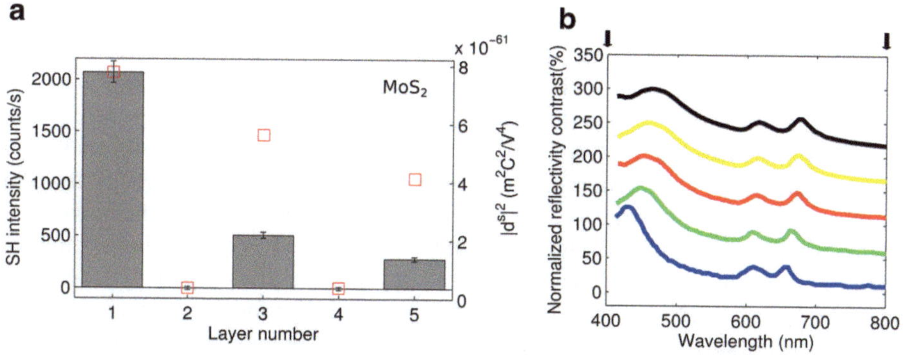

Fig. 6.4 (**a**) The layer dependence of SH intensity and surface nonlinear susceptibility for few-layer MoS_2 samples. The histogram shows the experimental results, with uncertainties indicated by the error bars. The *red squares* show the results of the model (described in the text) that has been applied for h-BN, but now also includes the effect of absorption of the fundamental and SH fields. (**b**) The relative change of reflectance of few-layer MoS_2 samples on a fused silica substrate. Successive spectra from the bottom to the top correspond to 1–5 layers. The spectra have been normalized for layer thickness. For clarity, the spectra have also been displaced vertically by 50 percentage units for each additional layer. The *arrows* indicate the wavelengths of fundamental and SH photons. (Based on Figure S1(a) in the Supplementary Material of reference [24])

without accounting for any change (symmetry lowering) in the inherent properties of the top monolayer of material, the effect typically considered in analysis of surface SHG [21]. Our simple model predicts the bulk SH response to be 0.20 % of that of a monolayer, comparable to the experimentally observed ratio.

6.6 SHG as an Optical Tool for the Characterization of Atomically Thin Layers

The SH probe has several useful properties for the characterizing few-layer MoS_2 and h-BN materials. First, the high contrast of SH intensities between even and odd numbers of layers can be used to supplement other optical characterization tools of layer number. The method is distinctive in providing strong contrast between single and bilayer samples. For MoS_2 with odd numbers of layers, even the exact layer number can be determined from the strong variation of SH intensity with layer number. Second, from the dependence of the SHG on the orientation of the crystal, we can determine the orientation of the crystallographic axes for odd-layer MoS_2 and h-BN. This purely optical method requires minimal sample preparation, is non-invasive, and imposes no special environmental requirements. Third, owing to the strong SH response for odd number of layers of MoS_2, the approach can be used for high throughput spatial mapping of domain orientation in polycrystalline samples, like those recently synthesized by chemical vapor deposition (CVD) [23, 34–37].

6.7 Conclusion

In conclusion, by applying optical second-harmonic generation spectroscopy, we have experimentally probed the symmetry properties of few-layer MoS_2 and h-BN. We verified that the few-layer samples retain the lattice symmetry they possess as part of the bulk material. The strong enhancement of SH intensity for odd numbers of layers compared to even numbers of layers reflects the broken inversion symmetry for samples of odd layer thickness and the recovery inversion symmetry for even layer thickness. The layer expected dependence of SH intensity from few-layer MoS_2 and h-BN was considered within a model accounting for wave propagation effects. Good agreement with the measured layer dependence was obtained for few-layer h-BN, but only the correct trend was predicted for few-layer MoS_2. The discrepancy between the model and experiment for MoS_2 indicates the importance of the material's evolving electronic structure with layer thickness as a result of interlayer electronic coupling. The orientational dependences of the SH intensity from MoS_2 and h-BN for odd layer thickness is consistent with the threefold rotational symmetry of the lattice and was demonstrated to be a useful tool for the determination of the orientation of the material's crystallographic axes.

References

1. Li, Y., Rao, Y., Mak, K.F., You, Y., Wang, S., Dean, C.R., Heinz, T.F.: Probing symmetry properties of few-layer MoS2 and h-BN by optical second-harmonic generation. Nano Lett. **13**, 3329–3333 (2013)
2. Castro, E.V., Novoselov, K.S., Morozov, S.V., Peres, N.M.R., dos Santos, J.M.B.L., Nilsson, J., Guinea, F., Geim, A.K., Neto, A.H.C.: Biased bilayer graphene: semiconductor with a gap tunable by the electric field effect. Phys. Rev. Lett. **99**, 216802 (2007)
3. Zhang, Y., Tang, T.-T., Girit, C., Hao, Z., Martin, M.C., Zettl, A., Crommie, M.F., Shen, Y.R., Wang, F.: Direct observation of a widely tunable bandgap in bilayer graphene. Nature **459**, 820–823 (2009)
4. Mak, K.F., Lui, C.H., Shan, J., Heinz, T.F.: Observation of an electric-field-induced band gap in bilayer graphene by infrared spectroscopy. Phys. Rev. Lett. **102**, 256405 (2009)
5. Cao, T., Wang, G., Han, W., Ye, H., Zhu, C., Shi, J., Niu, Q., Tan, P., Wang, E., Liu, B., Feng, J.: Valley-selective circular dichroism of monolayer molybdenum disulphide. Nat. Commun. **3**, 887 (2012)
6. Mak, K.F., He, K., Shan, J., Heinz, T.F.: Control of valley polarization in monolayer MoS2 by optical helicity. Nat. Nano **7**, 494–498 (2012)
7. Zeng, H., Dai, J., Yao, W., Xiao, D., Cui, X.: Valley polarization in MoS2 monolayers by optical pumping. Nat. Nano **7**, 490–493 (2012)
8. Duerloo, K.-A.N., Ong, M.T., Reed, E.J.: Intrinsic piezoelectricity in two-dimensional materials. J. Phys. Chem. Lett. **3**, 2871–2876 (2012)
9. Duerloo, K.-A.N., Reed, E.J.: Flexural electromechanical coupling: a nanoscale emergent property of boron nitride bilayers. Nano Lett. **13**, 1681–1686 (2013)
10. Michel, K.H., Verberck, B.: Theory of elastic and piezoelectric effects in two-dimensional hexagonal boron nitride. Phys. Rev. B **80**, 224301 (2009)
11. Michel, K.H., Verberck, B.: Phonon dispersions and piezoelectricity in bulk and multilayers of hexagonal boron nitride. Phys. Rev. B **83**, 115328 (2011)
12. Wu, W., Wang, L., Li, Y., Zhang, F., Lin, L., Niu, S., Chenet, D., Zhang, X., Hao, Y., Heinz, T.F., Hone, J., Wang, Z.L.: Piezoelectricity of single-atomic-layer MoS2 for energy conversion and piezotronics. Nature **514**, 470–474 (2014)
13. Zhu, H., Wang, Y., Xiao, J., Liu, M., Xiong, S., Wong, Z.J., Ye, Z., Ye, Y., Yin, X., Zhang, X.: Observation of piezoelectricity in free-standing monolayer MoS2. Nat. Nano **10**, 151–155 (2015)
14. Wilson, J.A., Yoffe, A.D.: The transition metal dichalcogenides discussion and interpretation of the observed optical, electrical and structural properties. Adv. Phys. **18**, 193–335 (1969)
15. Pease, R.: An X-ray study of boron nitride. Acta Crystallog. **5**, 356–361 (1952)
16. Novoselov, K.S., Jiang, D., Schedin, F., Booth, T.J., Khotkevich, V.V., Morozov, S.V., Geim, A.K.: Two-dimensional atomic crystals. Proc. Natl. Acad. Sci. U.S.A. **102**, 10451–10453 (2005)
17. Zeng, H., Liu, G.-B., Dai, J., Yan, Y., Zhu, B., He, R., Xie, L., Xu, S., Chen, X., Yao, W., Cui, X.: Optical signature of symmetry variations and spin-valley coupling in atomically thin tungsten dichalcogenides. Sci. Rep. **3**, 1608 (2013)
18. Malard, L.M., Alencar, T.V., Barboza, A.P.M., Mak, K.F., de Paula, A.M.: Observation of intense second harmonic generation from MoS2 atomic crystals. Phys. Rev. B **87**, 201401 (2013)
19. Shen, Y.R.: Optical second harmonic generation at interfaces. Ann. Rev. Phys. Chem. **40**, 327–350 (1989)
20. McGilp, J.F., Weaire, D.L., Patterson, C.H.: Epioptics: Linear and Nonlinear Optical Spectroscopy of Surfaces and Interfaces. Springer, Berlin (1995)
21. Reider, G.A., Heinz, T.F.: Second-order nonlinear optical effects at surfaces and interfaces: recent advances. In: Halevi, P. (ed.) Photonic Probes of Surfaces: Electromagnetic Waves, vol. 2. Elsevier, Amsterdam (1995) (Chapter 9)
22. Wagoner, G.A., Persans, P.D., Van Wagenen, E.A., Korenowski, G.M.: Second-harmonic generation in molybdenum disulfide. J. Opt. Soc. Am. B **15**, 1017–1021 (1998)

23. Kumar, N., Najmaei, S., Cui, Q., Ceballos, F., Ajayan, P.M., Lou, J., Zhao, H.: Second harmonic microscopy of monolayer MoS2. Phys. Rev. B **87**, 161403 (2013)
24. Mak, K.F., Lee, C., Hone, J., Shan, J., Heinz, T.F.: Atomically thin MoS2: a new direct-gap semiconductor. Phys. Rev. Lett. **105**, 136805 (2010)
25. Splendiani, A., Sun, L., Zhang, Y., Li, T., Kim, J., Chim, C.-Y., Galli, G., Wang, F.: Emerging photoluminescence in monolayer MoS2. Nano Lett. **10**, 1271–1275 (2010)
26. Gorbachev, R.V., Riaz, I., Nair, R.R., Jalil, R., Britnell, L., Belle, B.D., Hill, E.W., Novoselov, K.S., Watanabe, K., Taniguchi, T., Geim, A.K., Blake, P.: Hunting for monolayer boron nitride: optical and Raman signatures. Small **7**, 465–468 (2011)
27. Shen, Y.R.: The Principles of Nonlinear Optics. Wiley, New York (2003)
28. Stolz, W., Maan, J.C., Altarelli, M., Tapfer, L., Ploog, K.: Absorption spectroscopy on Ga0.47In0.53As/Al0.48In0.52As multi-quantum-well heterostructures. I. Excitonic transitions. Phys. Rev. B **36**, 4301–4309 (1987)
29. Murray Sargent, I., Scully, M.O.: Laser Physics. Perseus Books, New York (1978)
30. Bloembergen, N., Pershan, P.S.: Light waves at the boundary of nonlinear media. Phys. Rev. **128**, 606–622 (1962)
31. Hagimoto, K., Mito, A.: Determination of the second-order susceptibility of ammonium dihydrogen phosphate and alpha-quartz at 633 and 1064 nm. Appl. Opt. **34**, 8276–8282 (1995)
32. Beal, A.R., Hughes, H.P.: Kramers-Kronig analysis of the reflectivity spectra of 2H-MoS 2, 2H-MoSe 2 and 2H-MoTe 2. J. Phys. C **12**, 881 (1979)
33. Ishii, T., Sato, T.: Growth of single crystals of hexagonal boron nitride. J. Cryst. Growth **61**, 689–690 (1983)
34. Shi, Y., Zhou, W., Lu, A.-Y., Fang, W., Lee, Y.-H., Hsu, A.L., Kim, S.M., Kim, K.K., Yang, H.Y., Li, L.-J., Idrobo, J.-C., van der Kong, J.: Waals epitaxy of MoS2 layers using graphene as growth templates. Nano Lett. **12**, 2784–2791 (2012)
35. Lee, Y.-H., Zhang, X.-Q., Zhang, W., Chang, M.-T., Lin, C.-T., Chang, K.-D., Yu, Y.-C., Wang, J.T.-W., Chang, C.-S., Li, L.-J., Lin, T.-W.: Synthesis of large-area MoS2 atomic layers with chemical vapor deposition. Adv. Mater. **24**, 2320–2325 (2012)
36. Zhan, Y., Liu, Z., Najmaei, S., Ajayan, P.M., Lou, J.: Large-area vapor-phase growth and characterization of MoS2 atomic layers on a SiO2 substrate. Small **8**, 966–971 (2012)
37. Liu, K.-K., Zhang, W., Lee, Y.-H., Lin, Y.-C., Chang, M.-T., Su, C.-Y., Chang, C.-S., Li, H., Shi, Y., Zhang, H., Lai, C.-S., Li, L.-J.: Growth of large-area and highly crystalline MoS2 thin layers on insulating substrates. Nano Lett. **12**, 1538–1544 (2012)

Chapter 7
Valley Splitting and Polarization by Zeeman Effect in Monolayer $MoSe_2$

7.1 Introduction

Monolayer $MoSe_2$ features two inequivalent valleys in the Brillouin zone of its electronic structure. The broken inversion symmetry of the monolayer allows this valley degree of freedom to be selectively accessed by optical helicity, providing a unique platform to probe and manipulate the charge carriers in the two valleys [2–8]. Since the valleys are linked by time-reversal symmetry, they are energetically degenerate, while the magnetic moments of the corresponding valley states are of the same magnitude, but have opposite sign [2, 9, 10]. Coupling to the valley magnetic moments by a magnetic field thus provide an attractive, but as yet unexplored method of breaking the valley degeneracy [11, 12]. This presents new opportunities for the study of the fundamental physical properties of the valley electronic states, as well as for the development of new approaches to valleytronic control.

In this work, we experimentally investigate the ability of a perpendicular magnetic field to tune the valley energies in monolayer $MoSe_2$ by valley-resolved magneto-photoluminescence (magneto-PL) spectroscopy. *Lifting of the valley degeneracy* is demonstrated through the opposite energy shifts induced in the excitonic transitions in the two valleys by the magnetic field. The magnitude of this Zeeman shift, 0.12 meV/T, agrees with the predicted magnetic moments of the valley states. In the presence of a magnetic field, with split K and K' valleys, we create an *equilibrium valley polarization*, i.e., an imbalance in the charge distribution in the two valleys, by doping the sample. This behavior is revealed by the variation of the relative emission intensity of the charged and neutral excitons. Further, by comparing the direction of the energy shift of the conduction band and the relative intensity of the negatively charged exciton, we are able to clarify the valley configuration of these bright trion state. In addition, the doping dependent trion Zeeman shift reveals the modification to the many-body binding energy by the creation of valley polarization.

Y. Li, *Probing the Response of Two-Dimensional Crystals by Optical Spectroscopy*, Springer Theses, DOI 10.1007/978-3-319-25376-3_7

7.2 Experimental Setup for Valley Selective Magneto-PL Spectroscopy

$MoSe_2$ monolayers were prepared by mechanical exfoliation of bulk crystals on SiO_2 (300 nm)/Si substrates. Ti/Au contacts to the monolayer were fabricated using electron-beam lithography. With these electrical contacts, we could control the charge density of the sample by gating with respect to the Si substrate. The PL measurements were performed using excitation by a continuous-wave laser with a photon energy of 2.33 eV. The pump radiation was focused by an objective to a spot size of 3 μm on sample, with a power of ~100 μW. The reflected PL signal was collected by the same objective, and then filtered by a circular polarizer composed of a broadband quarter wave plate and a linear polarizer. The circularly polarized PL light was measured by a spectrometer equipped with a liquid-nitrogen cooled CCD detector. The PL measurements were carried out at a temperature of 10 K under a perpendicular magnetics field ranging from −10 T to +10 T.

7.3 Demonstration of Valleys Splitting and Polarization

In Fig. 7.1a, b, we present the magnetic field dependence of the PL spectrum for the right (σ_+) and left (σ_-) circularly polarized photons in the low-doping regime. We see prominent and well-separated features from emission of neutral (X^0) and charged (X^-) excitons [13–18]. The X^- peak lies ~30 meV below the X^0 peak, with an energy difference reflecting the trion binding energy. The emission energies of the neutral and charged excitons shift monotonically with magnetic field. The sign of the shift is reversed for photons of opposite circular polarization. Since the emitted photons with σ_+ and σ_- polarization states correspond to optical transitions from the different valleys [3–5], the fact that they shift in opposite directions is clear evidence that the K/K' valley degeneracy has been lifted, i.e., that valley splitting has been produced. The valley splitting is a direct consequence of the opposite magnetic moments associated with the bands in the two valleys, governed by the time reversal relation of the unperturbed valleys [6].

We extract the transition energy of the neutral exciton emission by fitting each spectral peak to a Lorentzian line shape. The resulting variation of the X^0 energy with the magnetic field is plotted in Fig. 7.1c, d. Using a linear fit, we obtain a rate of the Zeeman shift for the neutral exciton of −0.12 meV / T and +0.12 meV / T for the K and K' valleys, respectively. The experimental uncertainty of the extracted Zeeman slope is ±0.01 meV/T based on multiple measurements of the same sample.

To analyze the measured Zeeman shifts, let us recall the relevant band structure of $MoSe_2$ near the K/K' points. Both the conduction and valence bands, which have extrema at the K/K' points, are split by spin-orbit interactions [19–21]. For $MoSe_2$ the highest valence band and the lowest conduction band are predicted to have the same spin and to give rise to the observed bright exciton state. The other spin-split

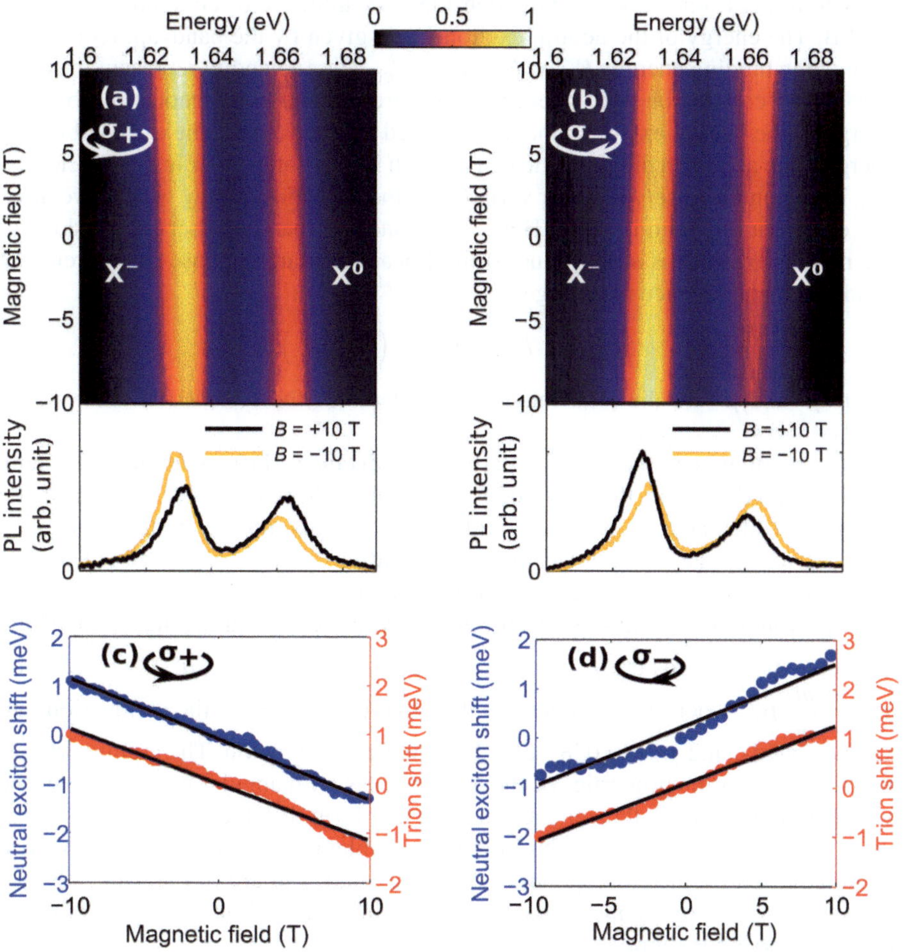

Fig. 7.1 (**a, b**) False color representation of the σ_+ and σ_- polarized PL spectra of monolayer $MoSe_2$ in the low-doping regime as a function of the strength of the applied perpendicular magnetic field. The spectra for fields of ±10 T are presented below the color map. The spectra are normalized to yield the same integrated intensity for the neutral and charged excitons, with a smooth background subtracted. (**c, d**) Extracted Zeeman shift of the neutral and trion energies for σ_+ and σ_- PL. The *solid lines* are linear fits to the experimental data

bands, separated by ~30 meV and ~200 meV, respectively, for the conduction and valence bands are not accessed in our PL measurements, which are conducted at liquid-helium temperature and at modest levels of sample doping. Since the relevant conduction and valence bands located are well described by a quadratic dispersion at the K and K' points [20], we analyze the magneto-optic response in a two-band effective mass framework throughout this chapter. The significant excitonic effects are considered through the introduction of binding energies for these states.

Within this framework, we now consider the influence of an applied magnetic field B. The energy of the neutral exciton E_{X0} is given by the bandgap, reduced by the exciton binding energy: $E_{X0} = E^c - E^v - E^b_{X0}$. Here E^c and E^v represent the energies of the conduction and valence band extreme, and $E_{X0}{}^b$ denotes the binding energy of the neutral exciton. Under an applied magnetic field, the exciton binding energies can acquire a diamagnetic shift, which is quadratic under in the B field [22, 23]. Based on the observed linear variation of the transition energy with B, we infer, however, that this term is small. Hence, we conclude that the observed PL shift for the neutral excitons is simply due to the linear Zeeman response of the relevant conduction and valence band edges:

$$E_{X0} = E^c - E^v = -\left(\mu^c - \mu^v\right)B,$$

where μ^c and μ^v are the total magnetic moments of the conduction and valence bands.

We can then compare our experimental Zeeman shifts with the magnetic moments predicted on the basis of a two-band tight-binding description of the electronic states near the K and K' valleys [2]. In this model, the total magnetic moment of a charge carrier arises from contributions of the atomic orbital (μ_l), carrier spin (μ_s), and of the delocalized Bloch wavefunction (μ_k), which we denote as the wavefunction contribution. In the K valley, the contributions to the magnetic moments of the bands are $\mu_s^c = \mu_s^v = -\mu_B$, $\mu_l^c = 0$, $\mu_l^v = -2\mu_B$, $\mu_k^c = -\frac{m_0}{m_*^c}\mu_B$, and $\mu_k^v = -\frac{m_0}{m_*^v}\mu_B$, where $m_*{}^c$ and $m_*{}^v$ are the effective masses of the conduction and valence bands, and $\mu_B = 0.058\,\text{meV}/\text{T}$ is the Bohr magneton. The atomic orbitals moments reflect the properties of the Mo *3d* states with angular moment $2\hbar$ or the valence band and 0 for the conduction band that form the electronic states at the K point. The wavefunction contributions are equivalent to a Bohr magneton for each band, adjusted by the corresponding effective mass, and are of the same sign for both bands. All of the contributions in the K' valley are by time-reversal symmetry, exactly opposite in sign.

In Fig. 7.2a, b, we illustrate the Zeeman shifts of the bands under the applied magnetic field, as well as the initial and final states for emission by a neutral exciton. Since interband optical transitions connect states in the conduction and valence bands with the same spin, the spin contribution to the Zeeman shifts cancel one another out in the transition energy. In the approximation of electron–hole symmetry, we have $m_*^c = m_*^v = m_*$ and $\mu_k^c = \mu_k^v = -\frac{m_0}{m_*}$. The wavefunction contributions to the Zeeman shifts then also cancel out in Zeeman shift of the transition energy [2]. For the K valley, we have accordingly a predicted Zeeman shift of the neutral exciton energy that depends only on magnetic moments of the Mo orbital angular momentum, namely, $\frac{E_{X0}}{B} = \mu^v - \mu^c = -2\mu_B = -0.116\,\text{meV}/\text{T}$. The rather precise agreement between this model and our experiment is fortuitous: The slightly larger hole mass obtained within a three-band tight binding model would, for example, contribute ~0.02 meV/T to the predicted Zeeman shift [20]. The sign change of the

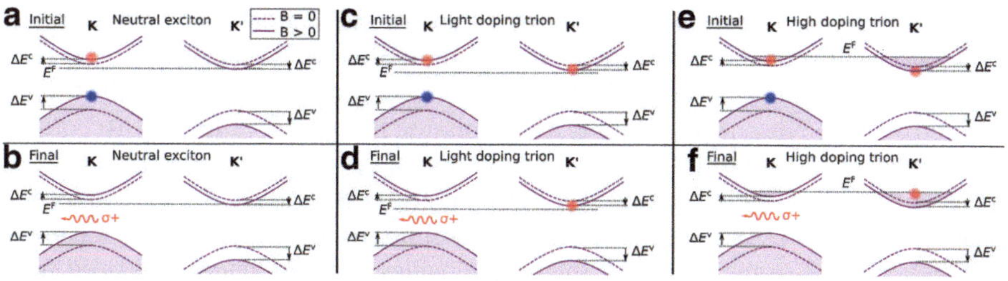

Fig. 7.2 Schematic of valley splitting and polarization from the Zeeman effect and the initial and final states of the neutral and charged exciton emission process. In the presence of an out-of-plane magnetic field, the conduction and valence bands shift according to their respective magnetic moments. The *dashed lines* denote the bands at B = 0 and the *solid lines* correspond to the bands for B > 0. (**a, b**) The initial and final states of the neutral exciton emission process in the low-doping regime. (**c, d**) The initial and final states of the charged exciton emission process in the low-doping regime. (**e, f**) The initial and final states of the trion emission process in the high-doping regime

Zeeman shift for the K' valley observed experimentally is, however, a robust theoretical prediction [10].

7.4 Determination of the Trion Configuration

Now we turn our focus to the trion emission. In the low-doping regime, our experiment yields Zeeman shifts for the negative trion of –0.12meV / T and +0.12meV / T for the K and K' valleys, respectively (Fig. 7.1c, d). Within experimental uncertainty, these values are the same as those found for neutral excitons. The trion emission energy can be expressed as $E_{X-} = E^c - E^v - E^b_{X0} - E^b_{X-}$, where E^b_{X-} denotes the binding energy of a trion, i.e., the energy needed to dissociate it into a neutral exciton and a free electron [17, 24]. The identical Zeeman slopes for X^0 and X^- emission is a consequence of the weak influence of the magnetic field on $E_{X0}{}^b$ and E^b_{X-} at low doping levels.

Considering the B-field induced splitting of the K and K' valley energies together with the finite electron doping of the sample, we expect the creation of an imbalance in the charge distribution in the two valleys, i.e., to the creation of static *valley polarization*. The presence of valley polarization is, in fact, clearly reflected in the relative intensities of the PL spectra of Fig. 7.1a. Since we detect *valley-specific* emission through the polarization selection, the relative PL intensity of charged and neutral excitons, I_{X-} / I_{X0} provides information about the presence of charge carriers within a given valley. To understand this phenomenon in more detail, let us consider the possible valley configurations of the trion species. A negative trion species contributing to emission in the K valley consists of an electron and hole in the K valley, together with an additional electron that could be located either in the K or K' valley. Because of the correlation between electron spin and valley, these nominally degenerate states have different electron spin configurations [6], the intravalley trion corresponding to identical electron spins and the intervalley trion corresponding to opposite electron spins. Exchange interactions then lead to energy differences between these two valley configurations of the trion [19].

To determine whether one of the two possible valley configurations for the trion is favored over the other, let us examine trion emission from the K valley (σ_+). According to the magnetic moments in the two-band model [2], a positive magnetic field induces an upshift of the K valley conduction band with respect to the K' valley conduction band (Fig. 7.2c, d). As a result, in thermal equilibrium more electrons populate the conduction band in the K' valley than the K valley. This produces enhanced trion emission for the case of intervalley trions. For the case of intravalley trions, the situation is exactly reversed and the K valley trion emission will be reduced.

Experimentally, we see that I_{X-} / I_{X0} for the K valley increases with magnetic field (Fig. 7.1a). We conclude that the trion emission originates predominantly from *intervalley* trions. This finding is compatible with recent theoretical prediction that the intervalley trion is ~6 meV more stable than the intravalley trion due to exchange interactions [19]. The intensity of the circularly polarized trion emission thus not only provides a direct signature of the creation of valley polarization, but also provides important information about the valley character of the trion species itself.

7.5 Zeeman Effect in the High Doping Regime

We have also studied the Zeeman effect in the regime of high carrier doping. Through electrostatic gating, we increase the charge density in the sample by $\sim 2.7 \times 10^{12}\ \mathrm{cm}^{-2}$, estimated from the gate capacitance and the applied voltage. Considering an electron band mass of 0.6 m_0 [20], this density of electrons would cause the Fermi energy E^f to lie 11 meV above the conduction band minimum. Note that in this regime, the finite (10 K) sample temperature can be neglected. The corresponding dependence on magnetic field of the circularly polarized PL spectra is shown in Fig. 7.3a, b. In this high-doping regime, the neutral exciton peak is very weak, and we only analyze shifts in the trion emission energy. We again find a linear Zeeman shift for the trion, with opposite behaviors in the two values. However, the rate of the trion Zeeman shift at high doping is $\mp 0.18\,\mathrm{meV}/\mathrm{T}$ for the K and K' valleys, respectively, compared with $\mp 0.12\,\mathrm{meV}/\mathrm{T}$ for the same quantity in the regime of low charge density.

We now consider the mechanism leading to the observed 50 % increase in the Zeeman shift of the trion in the high-doping regime compared to the low-doping regime. We attribute the changed behavior to the influence of the magnetic field on the trion binding energy E^b_{X-}. When the Fermi level is raised above the conduction band edge, E^b_{X-} acquires a term due to a state-filling effect, in which additional energy is required to move the extra bound electron from the conduction band edge back to the Fermi level when the trion is dissociated, i.e., $E^b_{X-}\left(E_f \geq E_c\right) = E^b_{X-}\left(E_f < E_c\right) + E_f - E_c$ [17, 24].

Figure 7.2e, f illustrates the Zeeman shifts of the bands, as well as the initial and final states of the trion emission process in the high-doping regime. The conduction

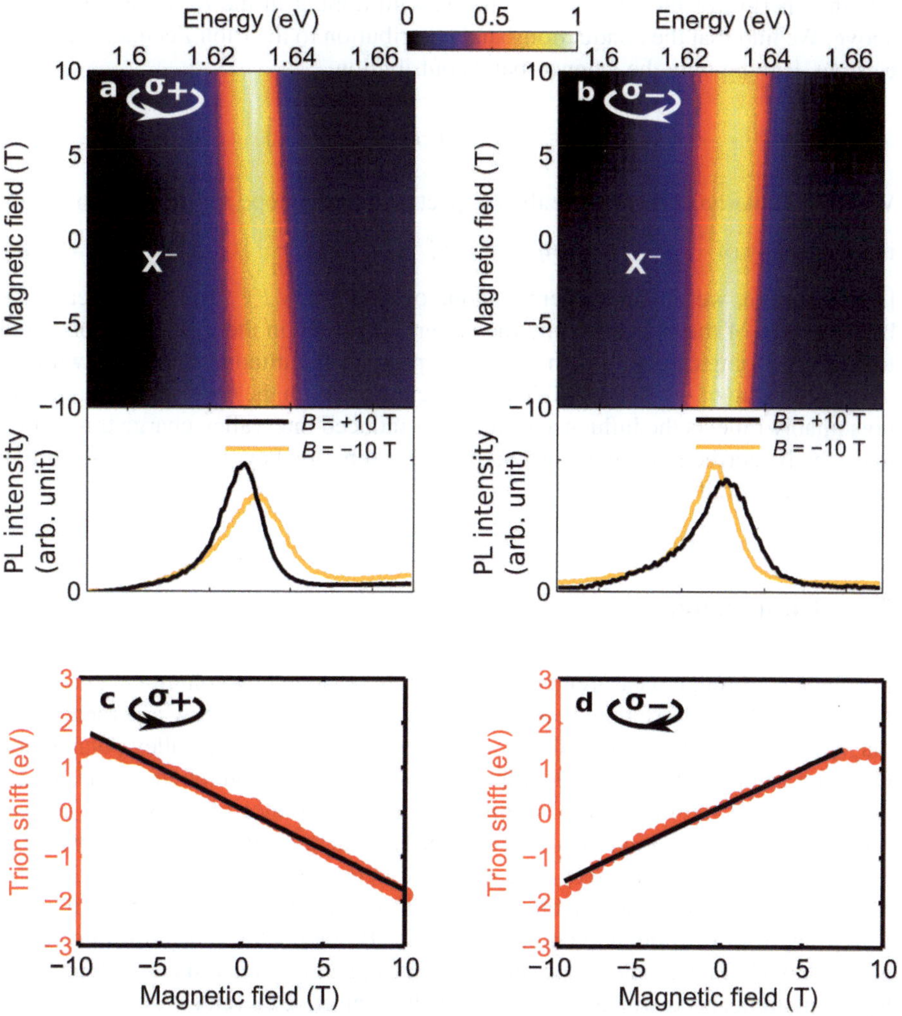

Fig. 7.3 Photoluminescence of monolayer $MoSe_2$ in the high-doping regime as a function of applied magnetic field. All information is as in Fig. 7.1, except only the negative trion feature is observable and no data are available for the neutral exciton

band splitting results in charge transfer from one valley to the other and the production of valley polarization, as described above. Based on the total magnetic moment of the conduction band in the two-band model, we deduce an electron transfer between the two valleys of $\Delta n / \Delta B \approx 3.9 \times 10^{10} \mathrm{cm}^{-2} / \mathrm{T}$ assuming an electron effective mass of $0.6\, m_0$ [20]. Despite the valley charge transfer process, the Fermi energy, we note, remains unchanged since the density of states in the two valleys is equal.

The Zeeman shift of the trion emission at high doping can be analyzed quantitatively based on the band alignments shown in the Fig. 7.2e, f and the variation of the binding energy for the (intervalley) trions with doping in the K' valley described above. We find that the conduction-band contribution to the trion Zeeman shift cancels out, leaving only the valence-band contribution:

$$\Delta E_{X-} = -\Delta E^{v} = \mu^{v} B, \quad \left(E^{f} > E^{c}\right)$$

Within the two-band model, we then expect a Zeeman slope for the trion emission in the K valley of $\frac{\Delta E_{X-}}{\Delta B} = \mu^{v} = \mu_{l}^{v} + \mu_{s}^{v} + \mu_{k}^{v} = -\left(2+1+\frac{m_0}{m^*}\right)\mu_B$. Using $m^* = 0.6\,m_0$ [20] as before, we obtain a Zeeman slope of $-0.27\,\mathrm{meV/T}$. This model captures both the sign of the trion shift and the larger magnitude in the trion shift compared to the low-doping case, although it does not provide quantitative agreement with the measured value of −0.18 meV/T. The difference between the experiment and model presumably reflects the influence of the field-induced intervalley charge transfer on many-body corrections to the trion transition energy beyond the trion binding energy considered above.

7.6 Conclusion

In summary, through measurements of circularly polarized photoluminescence from neutral and charged excitons in $MoSe_2$, we have demonstrated how a perpendicular magnetic field lifts the *K/K′* valley degeneracy and also creates a valley polarization in doped samples. We have identified the intervalley configuration as the lower-energy state for the trion and have discussed the importance of many-body corrections in the observed variation in the trion emission energy with doping. The controlled lifting of the valley degeneracy by a magnetic field and corresponding creation of a steady-state valley polarization open up exciting opportunities to probe and utilize the valley degree of freedom in monolayer TMDCs. By tuning the valley transition energies, it should, for example, be possible to create optically excited states with intervalley quantum beats, as well as steady-state valley selective currents.

Note added: The phenomenon of valley splitting has also been reported by other groups in monolayer $MoSe_2$ [25] and WSe_2 [26, 27]. Breaking of valley degeneracy by optical stark effect has also been demonstrated [28, 29].

References

1. Li, Y., Ludwig, J., Low, T., Chernikov, A., Cui, X., Arefe, G., Kim, Y.D., van der Zande, A.M., Rigosi, A., Hill, H.M., Kim, S.H., Hone, J., Li, Z., Smirnov, D., Heinz, T.F.: Valley splitting and polarization by the Zeeman effect in monolayer MoSe2. Phys. Rev. Lett. **113**, 266804 (2014)
2. Xu, X., Yao, W., Xiao, D., Heinz, T.F.: Spin and pseudospins in layered transition metal dichalcogenides. Nat. Phys. **10**, 343–350 (2014)
3. Mak, K.F., He, K., Shan, J., Heinz, T.F.: Control of valley polarization in monolayer MoS2 by optical helicity. Nat. Nano **7**, 494–498 (2012)

4. Zeng, H., Dai, J., Yao, W., Xiao, D., Cui, X.: Valley polarization in MoS2 monolayers by optical pumping. Nat. Nano **7**, 490–493 (2012)
5. Cao, T., Wang, G., Han, W., Ye, H., Zhu, C., Shi, J., Niu, Q., Tan, P., Wang, E., Liu, B., Feng, J.: Valley-selective circular dichroism of monolayer molybdenum disulphide. Nat. Commun. **3**, 887 (2012)
6. Xiao, D., Liu, G.-B., Feng, W., Xu, X., Yao, W.: Coupled spin and valley physics in monolayers of MoS2 and other group-VI dichalcogenides. Phys. Rev. Lett. **108**, 196802 (2012)
7. Mak, K.F., McGill, K.L., Park, J., McEuen, P.L.: The valley hall effect in MoS2 transistors. Science **344**, 1489–1492 (2014)
8. Sallen, G., Bouet, L., Marie, X., Wang, G., Zhu, C.R., Han, W.P., Lu, Y., Tan, P.H., Amand, T., Liu, B.L., Urbaszek, B.: Robust optical emission polarization in MoS2 monolayers through selective valley excitation. Phys. Rev. B **86**, 081301 (2012)
9. Yao, W., Xiao, D., Niu, Q.: Valley-dependent optoelectronics from inversion symmetry breaking. Phys. Rev. B **77**, 235406 (2008)
10. Xiao, D., Yao, W., Niu, Q.: Valley-contrasting physics in graphene: magnetic moment and topological transport. Phys. Rev. Lett. **99**, 236809 (2007)
11. Shkolnikov, Y.P., De Poortere, E.P., Tutuc, E., Shayegan, M.: Valley splitting of AlAs two-dimensional electrons in a perpendicular magnetic field. Phys. Rev. Lett. **89**, 226805 (2002)
12. Cai, T., Yang, S.A., Li, X., Zhang, F., Shi, J., Yao, W., Niu, Q.: Magnetic control of the valley degree of freedom of massive Dirac fermions with application to transition metal dichalcogenides. Phys. Rev. B **88**, 115140 (2013)
13. Mak, K.F., Lee, C., Hone, J., Shan, J., Heinz, T.F.: Atomically thin MoS2: a new direct-gap semiconductor. Phys. Rev. Lett. **105**, 136805 (2010)
14. Splendiani, A., Sun, L., Zhang, Y., Li, T., Kim, J., Chim, C.-Y., Galli, G., Wang, F.: Emerging photoluminescence in monolayer MoS2. Nano Lett. **10**, 1271–1275 (2010)
15. Chernikov, A., Berkelbach, T.C., Hill, H.M., Rigosi, A., Li, Y., Aslan, O.B., Reichman, D.R., Hybertsen, M.S., Heinz, T.F.: Exciton binding energy and nonhydrogenic Rydberg series in monolayer WS2. Phys. Rev. Lett. **113**, 076802 (2014)
16. He, K., Kumar, N., Zhao, L., Wang, Z., Mak, K.F., Zhao, H., Shan, J.: Tightly bound excitons in monolayer WSe2. Phys. Rev. Lett. **113**, 026803 (2014)
17. Mak, K.F., He, K., Lee, C., Lee, G.H., Hone, J., Heinz, T.F., Shan, J.: Tightly bound trions in monolayer MoS2. Nat. Mater. **12**, 207–211 (2013)
18. Ross, J.S., Wu, S., Yu, H., Ghimire, N.J., Jones, A.M., Aivazian, G., Yan, J., Mandrus, D.G., Xiao, D., Yao, W., Xu, X.: Electrical control of neutral and charged excitons in a monolayer semiconductor. Nat. Commun. **4**, 1474 (2013)
19. Yu, H., Liu, G.-B., Gong, P., Xu, X., Yao, W.: Dirac cones and Dirac saddle points of bright excitons in monolayer transition metal dichalcogenides. Nat. Commun. **5**, 3876 (2014)
20. Liu, G.-B., Shan, W.-Y., Yao, Y., Yao, W., Xiao, D.: Three-band tight-binding model for monolayers of group-VIB transition metal dichalcogenides. Phys. Rev. B **88**, 085433 (2013)
21. Kośmider, K., González, J.W., Fernández-Rossier, J.: Large spin splitting in the conduction band of transition metal dichalcogenide monolayers. Phys. Rev. B **88**, 245436 (2013)
22. Bayer, M., Walck, S.N., Reinecke, T.L., Forchel, A.: Exciton binding energies and diamagnetic shifts in semiconductor quantum wires and quantum dots. Phys. Rev. B **57**, 6584–6591 (1998)
23. Walck, S.N., Reinecke, T.L.: Exciton diamagnetic shift in semiconductor nanostructures. Phys. Rev. B **57**, 9088–9096 (1998)
24. Huard, V., Cox, R.T., Saminadayar, K., Arnoult, A., Tatarenko, S.: Bound states in optical absorption of semiconductor quantum wells containing a two-dimensional electron gas. Phys. Rev. Lett. **84**, 187–190 (2000)
25. MacNeill, D., Heikes, C., Mak, K.F., Anderson, Z., Kormányos, A., Zólyomi, V., Park, J., Ralph, D.C.: Breaking of valley degeneracy by magnetic field in monolayer MoSe2. Phys. Rev. Lett. **114**, 037401 (2015)
26. Aivazian, G., Gong, Z., Jones, A.M., Chu, R.-L., Yan, J., Mandrus, D.G., Zhang, C., Cobden, D., Yao, W., Xu, X.: Magnetic control of valley pseudospin in monolayer WSe2. Nat. Phys. **11**, 148–152 (2015)

27. Srivastava, A., Sidler, M., Allain, A.V., Lembke, D.S., Kis, A., Imamoglu, A.: Valley Zeeman effect in elementary optical excitations of monolayer WSe2. Nat. Phys. **11**, 141–147 (2015)
28. Kim, J., Hong, X., Jin, C., Shi, S.-F., Chang, C.-Y.S., Chiu, M.-H., Li, L.-J., Wang, F.: Ultrafast generation of pseudo-magnetic field for valley excitons in WSe2 monolayers. Science **346**, 1205–1208 (2014)
29. Sie, E.J., McIver, J.W., Lee, Y.-H., Fu, L., Kong, J., Gedik, N.: Valley-selective optical Stark effect in monolayer WS2. Nat. Mater. **14**, 290–294 (2015)

Chapter 8
Conclusion and Prospect

This thesis presents a series of studies of the response of two-dimensional crystals using optical spectroscopy. In the works conducted on graphene, we have probed the intrinsic electron–phonon and electron–electron interactions of ultraclean graphene on hexagonal boron nitride. We have demonstrated the strong out-of-plane field confinement of plasmons in graphene and enhanced electromagnetic interaction of graphene plasmon with adsorbed molecules and prototyped a graphene plasmon sensor. In the studies of transition metal dichalcogenides, we have measured the fundamental linear and nonlinear optical response of the materials, leading to further understandings of the excitonic properties and a useful nonlinear optical method to characterize the crystallographic orientation. In the magneto-optic study, we have investigated the ability to tune of the valley degree of freedom in monolayer transition metal dichalcogenides using a magnetic field. We demonstrated the valley splitting and polarization by the Zeeman effect of the valley states. Further, we discovered the important influence of valley polarization on the many-body binding energy of trion, as well as new fundamental insights into the trion electronic configuration.

To conclude this thesis, we provide a few prospects for future studies of these two-dimensional crystals. The ability to use a magnetic field to tune the valley degree of freedom in monolayer TMDCs has opened up new opportunities to understand and explore the valley physics in these materials. Possible directions include the creation of quantum beats in the excited coherent states in the two valleys and the generation of a steady state valley-spin current. The atomically thin TMDCs, with their strong electromagnetic response, may be useful for cavity quantum electrodynamics studies, where the light-matter interaction is further enhanced. With the strongly bound excitons, the ultrafast response of the excited excitons remains to be understood. How do excitons form? How do they relax? How do they interact with each other and other types of excitations? Last but not the least, combined effort in fundamental physical understanding, material growth, as well as device engineering will lead to the development of novel opto-electronic devices using atomically thin 2D materials.

Y. Li, *Probing the Response of Two-Dimensional Crystals by Optical Spectroscopy*, Springer Theses, DOI 10.1007/978-3-319-25376-3_8

Curriculum Vitae

Yilei Li
Department of Applied Physics
Stanford University
348 Via Pueblo Mall
Stanford, CA 94305, USA

Tel: (+1) 646-462-9695
E-mail: yileili@stanford.edu

EDUCATION

- Ph.D. in Physics, Columbia University, 2014 (Advisor: Prof. Tony F. Heinz)
- B.Eng. in Electrical Engineering, Hong Kong University of Science and Technology, 2010

PROFESSIONAL EXPERIENCE

- Postdoc in Prof. Tony Heinz group

  Stanford University, 2015 April–now
  Columbia University, 2014 October–2015 April

- Summer Research Internship in Dr. Phaedon Avouris group

  IBM Research, 2013 May–August

HONORS AND AWARDS

- Springer Prize for Outstanding Ph.D. Thesis, 2015
- Best Student Paper Award, Rank Prize Symposium, Grasmere, UK, 2014
- Townes Fellowship for Outstanding Research in Physics, Columbia University, 2014
- Academic Medal (highest distinction), Hong Kong University of Science and Technology, 2010
- University Scholarship, Hong Kong University of Science and Technology, 2006–2010

Y. Li, *Probing the Response of Two-Dimensional Crystals by Optical Spectroscopy*, Springer Theses, DOI 10.1007/978-3-319-25376-3

INVITED TALKS AT PROFESSIONAL MEETINGS

- "Magneto-optic study of monolayer transition metal dichalcogenides," Rank Prize Symposium on 2D Materials for Optoelectronics, Plasmonics and Photonics, Grasmere, UK, 2014
- "Probing the response of atomically thin transition metal dichalcogenides by optical spectroscopy," IGERT and Surface Interactions Symposium Series, Columbia University, New York, USA, 2014
- "Valley splitting and polarization by the Zeeman effect in monolayer $MoSe_2$," The 9th International Conference on Computational Physics, Singapore, 2015
- "2D materials beyond graphene: optical properties of atomically thin transition metal dichalcogenide crystals," Conference on Lasers and Electro-Optics (CLEO Europe), Munich, Germany, 2015 (Presented in place of Prof. T. F. Heinz)

CONTRIBUTED TALKS AT PROFESSIONAL MEETINGS

- "Second-harmonic generation from few-layer MoS_2 and h-BN," APS March Meeting, Baltimore, USA, 2013
- "Coupling of graphene plasmons to vibrations in molecules and solids," Office of Naval Research Funding Review, San Diego, 2014 (Presented in place of Prof. T. F. Heinz)
- "Valley splitting and polarization by the Zeeman effect in monolayer $MoSe_2$," APS March Meeting, San Antonio, USA, 2015
- "Linear and second-order nonlinear optical response of atomically thin transition metal dichalcogenide monolayers," The International Conference on Optics of Surfaces and Interfaces, Austin, USA, 2015

PUBLICATIONS COVERED IN THIS THESIS

1. Y. Li, I. Meric, K. F. Mak, L. M. Malard, L. Wang, C. Dean, P. Kim, J. Hone, K. L. Shepard, and T. F. Heinz, "Intrinsic doping dependence of Raman 2D mode in graphene: signatures of electron–electron interactions," (to be published)
2. Y. Li, H. Yan, D. B. Farmer, X. Meng, W. Zhu, R. M. Osgood, T. F. Heinz, and P. Avouris, "Graphene plasmon enhanced vibrational sensing of surface-adsorbed layers," Nano Lett. **14**, 1573–1577 (2014)
3. Y. Li, A. Chernikov, X. Zhang, A. Rigosi, H. Hill, A. M. van der Zande, D. A. Chenet, E. Shi, J. Hone, and T. F. Heinz, "Measurement of the optical dielectric function of monolayer transition metal dichalcogenides: $MoSe_2$, MoS_2, WSe_2, and WS_2," Phys. Rev. B **90**, 205422 (2014)
4. Y. Li, Y. Rao, K. F. Mak, Y. You, S. Wang, C. R. Dean, and T. F. Heinz, "Probing symmetry properties of few-layer MoS_2 and h-BN by optical second-harmonic generation," Nano Lett. **13**, 3329–3333 (2013)
5. Y. Li, J. Ludwig, T. Low, A. Chernikov, X. Cui, G. Arefe, Y. D. Kim, A. M. van der Zande, A. Rigosi, H. Hill, J. Hone, Z. Li, D. Smirnov, and T. F. Heinz, "Valley splitting and polarization by Zeeman effect in monolayer $MoSe_2$," Phys. Rev. Lett. **113**, 266804 (2014)

PUBLICATIONS NOT COVERED IN THIS THESIS

6. J. Stähler, O. T. Hofmann, P. Rinke, S. Blumstengel, F. Henneberger, Y. Li, and T. F. Heinz, "Raman study of 2,7-bis(biphenyl-4-yl-)2′,7′-ditertbutyl-9,9′-spirobifluorene adsorbed on oxide surfaces," Chem. Phys. Lett. **584**, 74–78 (2013)
7. X. Yin, Y. Li, F. Ke, H. Zhao, L. Gan, R. Zhao, T. F. Heinz, and Z. Hu, "Evolution of the Raman spectrum of graphene grown on copper upon oxidation of the substrate," Nano Res. doi:10.1007/s12274-014-0521-0 (2014)
8. A. Chernikov, T. C. Berkelbach, H. M. Hill, A. Rigosi, Y. Li, M. S. Hybertsen, D. R. Reichman, and T. F. Heinz, "Exciton binding energy and Nonhydrogenic Rydberg series in monolayer WS_2," Phys. Rev. Lett. **113**, 076802 (2014)
9. C. H. Lee, G. H. Lee, A. M. van der Zande, W. Chen, Y. Li, M. Han, X. Cui, G. Arefe, C. Nuckolls, T. F. Heinz, J. Guo, J. Hone, and P. Kim, "Atomically thin pn junctions with van der Waals heterointerfaces," Nat. Nanotech. **9**, 676–681 (2014)
10. W. Wu, L. Wang, Y. Li, F. Zhang, L. Lin, S. Niu, D. Chenet, X. Zhang, Y. Hao, T. F. Heinz, J. Hone, and Z. L. Wang, "Piezoelectricity of single-atomic-layer MoS_2 for energy conversion and piezotronics," Nature. **514**, 470–474 (2014)

MIX
Papier aus verantwortungsvollen Quellen
Paper from responsible sources
FSC® C105338

If you have any concerns about our products,
you can contact us on
ProductSafety@springernature.com

In case Publisher is established outside the EU,
the EU authorized representative is:
Springer Nature Customer Service Center GmbH
Europaplatz 3, 69115 Heidelberg, Germany

Printed by Libri Plureos GmbH
in Hamburg, Germany